SCIENCE AND TECHNOLOGY IN

MEDIEVAL EUROPEAN LIFE

SCIENCE AND TECHNOLOGY IN

MEDIEVAL EUROPEAN LIFE

JEFFREY R. WIGELSWORTH

The Greenwood Press "Daily Life Through History" Series
Science and Technology in Everyday Life

GREENWOOD PRESS
Westport, Connecticut•London

Library of Congress Cataloging-in-Publication Data

Wigelsworth, Jeffrey R.
    Science and technology in medieval European life / Jeffrey R. Wigelsworth.
        p. cm. — (The Greenwood Press daily life through history series, science
and technology in everyday life, ISSN 1080–4749)
    Includes bibliographical references and index.
    ISBN 0–313–33754–3
    1. Technology—Europe—History. 2. Science—Europe—History. 3. Civiliza-
tion, Medieval. I. Title.
    T26.A1W54 2006
    609.4'0902—dc22   2006021733

British Library Cataloguing in Publication Data is available.

Library of Congress Catalog Card Number: 2006021733
ISBN: 0–313–33754–3
ISSN: 1080–4749

First published in 2006

Greenwood Press, 88 Post Road West, Westport, CT 06881
An imprint of Greenwood Publishing Group, Inc.
www.greenwood.com

Printed in the United States of America

The paper used in this book complies with the
Permanent Paper Standard issued by the National
Information Standards Organization (Z39.48–1984).

10   9   8   7   6   5   4   3   2   1

CONTENTS

PREFACE

This book is meant to be a reference work used by both college students and any other interested readers. It is also an attempt to shine some much-needed light on the so-called Dark Ages in European history. Ever since Renaissance scholars celebrated their own achievements while simultaneously looking down their collective noses at the learning of those who came before them, the medieval era in Europe (roughly 500–1500) has been written off as a time when not much important happened intellectually and certainly no lasting science and technology were developed. Even among professional historians, there seems to be a tendency to treat medieval science and technology not as things of merit in themselves but rather as part of a search for what they might have contributed to more important epochs such as the Scientific Revolution in the seventeenth century and the Industrial Revolution of the late eighteenth and early nineteenth centuries. My hope is that readers will leave this book with a different impression.

Today, science and technology really capture the public imagination only when they impact daily life, and the same was true in medieval Europe. To illustrate various (though by no means all) engagements between science and technology with everyday life is the overall goal of this book. I chose what seemed to me aspects of science and technology that would likely have most directly impacted daily life in medieval Europe. Because this is a reference work that provides many examples of everyday science and technology, it does not have to be read cover to cover — though it certainly could be. Rather, I expect that most readers will pick one topic and focus on it. With this readership in mind, I have divided the book into thematic chapters containing subsections, each addressing a particular topic encompassed by that thematic grouping. For example, eyeglasses and papermaking are found in Chapter 4, on communication. Not all strata of society would have been affected by all the technologies and theories discussed in this book, but, collectively, daily life for Europeans as a whole changed even if individual lives did not.

After an introductory chapter, which paints a picture of medieval Europe that provides broad-brushstroke context for the era's science and technology, the book proceeds with a brief historical outline that contains a schematic narrative of the major scientific and technological achievements of medieval persons. This is followed by a chronology. Chapter 1 discusses the technology and science of earning a living, whether by farming or banking. Chapter 2 details the process of building castles and churches—two of the most lasting reminders of medieval know-how. Chapter 3 outlines the various methods of transportation of both people and products. Chapter 4 illustrates how medieval Europeans communicated in an age before email and cell phones. Chapter 5 reveals the heavy debt that medieval exploration owed to science and technology. Chapter 6 focuses on the most destructive aspect of medieval science and technology: warfare. After battles, physicians were often required; medicine both scholarly and folk is discussed in detail in Chapter 7. Chapter 8 has time as its topic, and readers learn how medieval Europeans used a variety of instruments to determine time intervals both short and long. Science and religion are the topic of Chapter 9, which reveals that sharp distinctions between the two are modern inventions and would not have been recognized in medieval daily life. The book ends with a lengthy list of suggested reading, and a glossary of historically specific terms. These features will be useful for students conducting research for their courses or anyone interested in learning more about topics presented in the book.

Throughout this book I have emphasized readability and accessibility to as wide an audience as possible. To this end, I have not paid strict attention to medieval word usage when it came to designating people "scientists" or "natural philosophers" (what medieval thinkers would have called scientists). Where it made a difference in meaning, such as in the chapter on science and religion, I have labeled people "natural philosophers" and their practice "natural philosophy" rather than "science" because this best captures their motives and purpose. This is explained in relevant sections within the text. In other instances, I have referred to "science" where, for the sake of clarity, this term seems best able to convey a medieval intellectual activity to modern readers. This is also the case when people are called "architects," which was a different profession from what it is now, but for expediency I have used it in a more modern sense while pointing out medieval uniqueness where it seemed prudent. In other cases, medieval word usage is maintained. Whereas we use "doctor" and "physician" somewhat interchangeably, medieval Europeans did not. A "physician" was someone who had attended university and obtained a M.D.; a doctor was what we would call a person with a Ph.D., not necessarily a medical doctor. Thus, I have followed this medieval convention throughout the text.

Since this is a synthetic work that brings together the latest research in an accessible form, I am greatly indebted to those historians whose work I have used. I now know what Isaac Newton meant when he told Robert Hooke in 1676: "If I have seen further it is by standing on the shoulders of giants." The list of Further Reading makes my debt clear, but I should like to offer my thanks to those scholars here. On a more personal note, I thank Kevin Downing, who first suggested that I undertake this project and offered support and criticism throughout. Linda Tweddell was an excellent editor. Maggie Osler and Larry Stewart were my teachers and between them, in their own different ways, opened my eyes to the history of science. For that I can only say thank you. This book was written at the Calgary Institute for the Humanities, an underappreciated treasure at the University of Calgary. I offer my thanks to Wayne McCready, the director; Denise Hamel, the administrator; and the other Fellows for providing an enjoyable and stimulating environment in which to work. My parents have always supported my studies of the past, even when they were not sure anyone else would be interested. For that, I am grateful. My wife, Alison (an accomplished scholar herself), proved an excellent critic and editor; her name deserves a place on the cover next to mine. She was encouraging and patient during this entire process. One could not ask for a more loving wife or a better friend.

INTRODUCTION

The Middle Ages or the medieval period in European history is generally viewed as that period of time from the fall of the Roman Empire until the peak of the Renaissance and the beginning of the Reformation—in chronological terms, roughly 500 C.E. to 1500 C.E. Within this broad classification are smaller divisions: early medieval (500–1000); high medieval (1000–1300); and late medieval (1300–1500). These dates are permeable, and not all scholars agree on them. Far from being the "Dark Ages," as the period was described by Renaissance thinkers who viewed it as an age of intellectual stagnation situated between the lost riches of classical antiquity and the rebirth of knowledge during the Renaissance, the medieval period in Europe produced much in the way of science, as well as lasting reminders of its technological know-how. Nonetheless, the terms used to describe this thousand-year period of history reflect the contempt for it that has lasted from the Renaissance. "Dark Ages" is the most pejorative of these in that it suggests a time in which a shadow of ignorance covered Europe. Similarly, "Middle Ages" reflects a view that this time sits between two better eras and that the Middle Ages served as something of a holding pattern in European history. The characterization of the time as "medieval," a term that has come to be synonymous with the period 500–1500, has its origins in a sense that the time was unenlightened. Thus, once more this millennium is named for something negative. The problem of identifying this parcel of history without immediately denigrating it has led some scholars to suggest abandoning these terms entirely. The historians Roger French and Andrew Cunningham have offered "Troubadourean," after the composers/singers, as a way to avoid all the latent assumptions present in past labels. However, this has yet to catch on with other scholars. What has changed dramatically, even if the names remain the same, is the view of medieval Europe (the term used in this book) as one that produced no lasting or important intellectual achievements. This book continues the endeavor of reclaiming the ignored impact of this era by presenting the considerable accomplishments of medieval Europe in the realms of science and technology and how these impacted daily life. The purpose of this introductory chapter is to paint a picture of medieval Europe that allows the science and technology depicted in the rest of book to be set into their historical contexts: social, political, and religious. It will become apparent that this was an age of dynamic change.

Following the deposition of the emperor Romulus Augustulus, in 476 C.E., the date traditionally accepted as the final fall of the Roman Empire, western areas of Europe became peopled by various Germanic groups or tribes. Visigoths controlled Spain, Vandals held northern Africa, Lombards ran Italy, and groups of Angles and Saxons vied for Britain. It was a time of uncertainty and constant warfare as competing factions battled for territory and prestige. Christianity, having been established as the official religion of Rome, continued as a cultural and spiritual force—albeit in a more localized sense rather than as an international organization radiating from Rome and the pope. During these decades and centuries, the intellectual and cultural splendor of the Roman Empire was somewhat forgotten. This happened not because it was a dark age when people were less intelligent or creative but rather because intellectual energies were directed toward survival and the acquisition of everyday necessities such as food and shelter, and these activities took precedence over poetry and oratory. In some places, learning did continue. Ireland, for example, produced celebrated works of Christian scholarship.

After the Frankish king Clovis (466–511) conquered modern France (then called Gaul), a sense of unity came to much of Europe. What is more, Clovis was Catholic and so too became his kingdom. Catholicism played such a major role in the Frankish kingdom that the Catholic Church invested subsequent kings. Earning a living during this era was difficult. Crop yields were poor, and the economy ran on barter rather than coin. After years of dynastic struggle for parcels land that became less and less sizable as members of each successive generation divided their father's land, the Franks were again united by Charles Martel around 732. Charles's son, Pepin the Short, with the backing of the papal office, was crowned king of the Franks in 752. Pepin's dynasty is known not for him but rather for his most famous descendant, Charlemagne (ca. 742–814); hence the term Carolingian. Charlemagne was a warrior, but he was also deeply Catholic and believed that he was reuniting the former Roman Empire under a Catholic sign; he would do so by force if necessary. The Carolingian Empire ran by the delegation of power to a system of dukes, counts, and bishops who acted in Charlemagne's name. Moreover, Charlemagne acted through the organization of the Catholic Church to ensure order. It was under Charlemagne that Catholicism adopted the parish system with local priests subordinate to their bishops and bishops subordinate to the pope. When the pope crowned Charlemagne emperor, in 800, it was symbolic recognition of what was already acknowledged: Charlemagne was at the helm of the Catholic Church. His reign is also known as the Carolingian Renaissance because Charlemagne promoted learning, libraries, and literature. He also established schools for future clergy so that Church officials would not be ignorant.

After the death of Charlemagne, his empire did not survive. Warring groups from various locations beset Europe, and whatever stability had been gained was lost. Muslim pirates sailed from northern Africa in search of riches, and Viking warriors came down from the Scandinavian nations. Collectively, this age is remembered for the invasions of the ninth and tenth centuries. These raids brought anarchy to Europe and fostered an everyone-for-himself attitude. During this time, a new social and political organization structure known as feudalism came into use. In its simplest form, feudalism functioned as a defensive social system based on obligations that ran from the highest to the lowest strata of society. This is discussed at length in Chapter 1. With feudalism came the technological achievements of castles and mounted knights in armor. Feudalism spread throughout Europe and characterizes the tenth and eleventh century until the end of what seemed like constant invasions finally came. What did survive out of feudalism was a notion of strong social hierarchy with monarchs at the top and various nobles below them; the same was also true of the Catholic Church, which was organized from the pope down in straight lines of expected obedience.

Heirs of Germanic tribes still vied for European kingdoms into the eleventh century. In 1066, the famed William the Conqueror invaded England and was victorious over the English king, Harold. Both William, a Norman, and Harold, a Saxon, were replaying centuries-old tribal conflicts over which group had rightful claim to certain lands. William, however, even though he was king of England and administered a well-run kingdom, still owned Normandy in France and as such was a subject of the French king. This situation would lead to the Hundred Years War between France and England, in which English kings sought autonomy in their lands and perhaps the French throne. Monarchs in France did not gain the same level of administrative control in their kingdom as existed in England until the twelfth century, and even then the land under their purview represented a fraction of the modern nation. Meanwhile, in England, the Magna Carta (1215) guaranteed certain rights to nobles but nothing for the rest of the populace.

The era of raids, discussed earlier, dramatically affected the Catholic Church, whose priests sought protection under their local nobles.

These secular rulers exerted tremendous influence and control over the running of Church institutions, such as monasteries, in their territories. Bishops were often second sons of nobles, who increasingly saw administration of the Church as their right. (First sons held their father's noble title and land.) Questions of who actually ran the Catholic Church—pope or prince—came to the fore of medieval debate. Not surprisingly, Church prestige fell dramatically. The Church attempted some rejuvenation of its image, for example by founding the Cluny monastery, which emphasized piety and complete devotion to God. Not wanting to be completely beholden to secular rulers who, if not directly appointed by monarchs, certainly had to be approved by them, the papacy under Pope Nicholas II established the precedent that future popes would be elected by the College of Cardinals; the first election of this type occurred in 1061. This left appointment of bishops an unresolved issue. Both the Church and princes claimed the right to appoint bishops, and this problem led to the Investiture Controversy, which would not be solved until 1122 with an agreement that the Church would grant the title of bishop and secular rulers would provide the land for the bishopric. The division of power between Church and monarch, and the question of which was superior, would be a pressing concern for much of the medieval era and beyond.

Despite this uneasy division of power, Europe collectively saw itself as Christendom and as being charged with protecting and propagating the Christian faith. This perceived duty is evident in the many Crusades sent to secure Jerusalem from Islamic Turks. When Pope Urban II first issued the call for a crusade in 1095, he was answered by a ragtag group of peasants led by a charismatic outcast called Peter the Hermit. Not surprisingly, Christian enthusiasm was not a match for the well-trained Turkish soldiers. A more militarily competent expedition was successful and captured Jerusalem in July 1099. Security in the Holy Land was important, as was caring for the fallen, and two new military orders, soldiers for Christ, were created during the Crusades: the Knights of the Hospital of St. John and the Knights Templars. Future crusades to reclaim land that had been gained and lost in a series of battles would ultimately prove a failure, with the low point coming during the Fourth Crusade (1202–1204). When crusading armies could not pay Venice for their passage across the Mediterranean, they were co-opted into attacking the Christian city of Constantinople, which was involved in a trade dispute with Venice.

Between the twelfth and thirteenth centuries, while crusading armies conducted warfare in the name of God, medieval scholars and theologians were engaged in their own battles. Beginning near the middle of the eleventh century, questions were raised about the truth of the transubstantiation miracle, in which the bread and wine of Communion are transformed into the body and blood of Christ, even though their outer appearance remains the same. For an institution like the Catholic Church that had not really been challenged in matters of theology since the days of the

Church fathers, this was a serious problem. The issue was the proper use (if any) of reasoned arguments in theology. Some theologians argued that reason (or philosophy) had no place in evaluating articles of theology: one needed only faith. Other theologians countered that reason might serve theology and religion very well. Still others applied their passion for reason to engagement with classical literature from Roman antiquity and initiated what scholars refer to as the Renaissance of the Twelfth Century because of its similarity to the more famous Renaissance that occurred in the fourteenth and fifteenth centuries. Aside from locating a proper relationship between reason and faith on a hierarchy of intellectual pursuits, the question of universals ensured that solutions would not be quickly found. This philosophical dispute centered on a deceptively simple issue: are concepts or attributes truly real? For example, does "coldness" exist independent of something that is "cold"? One group of thinkers, called "Realists," argued that such things do exist as separate entities. On the other side of the issue, "Nominalists" countered that "coldness" is nothing more than a name, a linguistic identification, with no real existence of its own. Disputes of this sort would remain common throughout the medieval period and would continue to intrigue modern philosophers and theologians. With the rise of the universities, also in the thirteenth century, intellectual investigations only increased.

The fourteenth century in Europe was one of both disaster and increased expectations. Disaster hit first. Between 1315 and 1317, northern parts of Europe experienced a devastating famine, known to historians as the Great Famine. Southern parts of Europe coped with their own famine during 1339–1340. The proximate causes of these issues was overpopulation and cooling climatic changes that ensured that not enough food would be available. Rates of procreation dropped precipitously during the famine, and those who did survive childhood during this time were chronically malnourished and unable to defend against the major disaster right around the corner. Arriving on Italian merchant ships returning from Asia and traveling throughout Europe along established trading routes, the Black Death, or bubonic plague, ravaged all of Europe between 1347 and 1351. Estimates of the number of deaths caused by the plague range from one-third to one-half of the entire population of Europe. At its pinnacle, there were not enough living to bury the dead. Every cloud has a silver lining, however, and those who did survive the plague benefited from the ordeal. Peasants were helped by a radical drop in food prices, and land values declined, too. Survivors from higher social strata inherited great quantities of land, and laborers demanded higher wages. With fewer workers competing for wages and demand for labor as high as it had been prior to 1347, those at the bottom of society reaped a financial windfall. Aristocratic landowners throughout Europe initiated legislation to curb rising wages and restrict the movement of peasants. The Statute of Laborers, issued in England in 1351, is a good example of this trend.

Peasants reacted harshly to this sort of law, and both France and England witnessed revolts between 1350 and 1360. In the case of Florence, the rebellions would last into the 1380s.

Recovering from the fourteenth century was the order of the day for much of the fifteenth. During this time, monarchies across Europe developed somewhat modern bureaucracies and administrations based on regular systems of taxation and defense conducted with salaried soldiers. Gone forever were the personal feudal obligations of duty and knightly service. Mercenaries and standing armies replaced these products of a bygone era. States ran on money to a greater extent than ever before. Employees of the crown who served the royal whim ensured their continued place by streamlining tax collection (thereby increasing the volume of money received) and found new sources of royal income. Requests for money from an established noble order that was used to doing pretty much as it pleased in the countryside created inevitable clashes that had to be resolved in newly created courts. Systems of civil law unique to each nation, based in part on Roman law, came to supplant Church canon law in many cases. Still, the system of state administration of this later medieval time was drastically different from our modern government. The concerns of monarchical dynasties and the requirement of religious uniformity dictated state policy to a greater degree than did anything resembling national interest. Weak monastic dynasties that could not keep their nobles in line fell, and others rose very quickly during the fourteen and early fifteenth centuries.

There was no monolithic medieval era or Middle Ages. The period from roughly 500 to 1500 was one of dynamic change and evolution. Styles of government came and went, control of religion was a hotly contested issue, and the role of lower-ranking people, such as peasants in their own kingdoms, was a worrying concern for the elite of society that believed "innovation" or "change" to be four-letter words. This same pattern of evolution and development is also seen in the era's science and technology.

HISTORICAL OUTLINE

Medieval Europe is usually agreed to have begun after the fall of the Roman Empire and to have lasted until the peak of the Renaissance and the eve of the Reformation in Europe, or approximately from 500 to 1500. During this time, many advances and innovations in science and technology occurred. While this brief outline of events cannot hope to cover them all, a hint of the invention and genius evident during this era is meant to whet the appetite for the detailed accounts found in the body of this book.

The first major figure in the history of science in western medieval Europe was St. Augustine (d. 430 C.E.), who took a great interest in the writings of the Greek philosopher Plato. Augustine argued that knowledge of nature gained through science was an acceptable activity for Christians so long as they remembered that the Bible was always superior and that science was to be a handmaiden to theology. Plato (d. 347 B.C.E.) was not the only ancient thinker who influenced medieval studies of nature. The writings of his student, Aristotle (d. 322 B.C.E.), would form much of the foundation for later medieval scholarship. One of the earliest translators of Aristotle from Greek into Latin was the Roman polymath Boethius (d. 524 C.E.), whose translations were the only version of Aristotle available in Europe until around 1200. Christianity provided a framework for doing science, and problems of the Christian faith were often the impetus for science. A case in point was the effort to determine the perpetual date of Easter, which involved mathematical astronomy and knowledge of calendars. An English monk named the Venerable Bede settled this issue in the early eighth century. Mathematics required numbers, and these were depicted as Roman numerals for much of the medieval era, but not among Arabic scholars, who employed what we call Arabic numerals (our modern number system) as early as 760. Arabic numerals would not be used commonly in Christian Europe until around 1200.

Paper made from pulp (initially cloth and some wood) was first made in China, and by 793 factories in Baghdad were producing great quantities of inexpensive paper. This ability to manufacture paper may explain why by ca. 900 Baghdad could boast of having more than 30 public libraries, a number that would not be exceeded in western Europe for generations. Paper did not arrive in Europe until between 900 and 1000; prior to this time, Europeans wrote on sheets of vellum or parchment that were fashioned from the skins of animals. This period was not without innovation in Europe, however, because the years around 900 witnessed the development of three important pieces of farming technology that fundamentally altered agriculture: the wheeled plow, iron horseshoes, and the horse collar.

The early eleventh century was a time of great accomplishment in medieval science and technology. Muslim scholars had translated nearly the entire body of Greek writings on science and medicine and began studying them and composing commentaries. Two of the most important Islamic people in this enterprise were Avicenna (a.k.a. Ibn Sina) (980–1037) and Averroes (a.k.a. Ibn Rushd) (1126–1198), whose writings, once translated from Arabic into Latin, would provide much intellectual stimulation in Christian circles and ensured Averroes's reputation as "the commentator." At the same time, Christian churches were taking on a new appearance that is known as Romanesque design. After William the Conqueror secured England in 1066, he wished to know the riches contained in the country and counted them in the Domesday Book, completed in 1086. Recorded are more than 6,000 waterwheels, which signals the degree to which water power had established itself as essential to the medieval economy and society. Windmills soon followed.

By the twelfth century, Romanesque churches, with their trademark thick stone walls, had given way to the grandeur of Gothic design, with its emphasis on high walls made more of glass than of stone. In the 1140s, knowledge of magnets became more common in Europe, which allowed for the development of compasses, around 1180, an essential aid to navigation on the high seas. During the same era, the trebuchet, a new terror weapon, was used for the first time. An offset catapult, the trebuchet dominated European warfare until cannon superseded it beginning around 1324. Establishment of silk weaving in Italy around 1150 meant that this luxury item no longer had to be imported from the East and dramatically reduced its cost, although it was still affordable only to the very elite of society. The middle of the twelfth century also brought the first European university, founded at Bologna, Italy, in 1158. Construction of the famed

London Bridge began in 1176 and would not be completed until 1209. The technology of building projects also filtered down to lesser structures. The first mention of a chimney in England occurred in 1185. While often taken for granted today, a chimney was a major innovation that permitted the use of fires for warmth without filling the room with smoke.

The thirteenth century was perhaps the most eventful for medieval science and technology. By 1200, Europe's scholars had translated from Arabic and into Latin the bulk of Aristotle's writings. The availability of this "new" philosophy occurred at around the same time that universities began to flourish, and the two events united in a new form of scholarship that would shape intellectual inquiry, including what we call science, for the next 500 years. Scholasticism, as the activity was known, involved posing questions of a philosophical bent—such as "could a vacuum exist?"—and seeking answers and opinion within an Aristotelian text or in the work of an accepted Aristotelian authority such as Averroes. The recovery and translation of ancient authors also sparked growth in medical education, centered early at Bologna and based on the writings of the Roman physician Galen, who would set the medical agenda in Europe for generations. Modern-looking banks emerged also around 1200 and forever changed business practices by allowing notes made on paper to be the currency of commerce rather than bags of coin. Closely related to banking and bookkeeping was the introduction of the number zero, which was a difficult concept for medieval thinkers because it was both a number and nothing. For those looking to stay warm on cold nights, the invention of buttons (also around 1200) permitted clothes to be bound up against the weather. In 1230, the first depiction of a wheelbarrow is seen, and a new age of transportation dawned. While this may seem a second-rate technology, laborers, who formerly had to carry their load in their arms or on their backs, would have disagreed. During 1257, the famed Dominican scholar St. Thomas Aquinas began teaching his mix of Aristotelian philosophy and Christian theology to great acclaim. His work reached its pinnacle in 1265 with the publication of *Summa contra Gentiles,* a kind of manual for showing how knowledge of science (what Aquinas would have referred to as natural philosophy) was a valuable weapon to use in the conversion of pagans to Christianity. The incorporation of Aristotelean theory into university curricula and its use by writers like Aquinas did not sit well with some authorities, and in 1277 the Bishop of Paris issued the famous condemnations against Aristotle. Although these prohibitions would be toothless in later years, their initial result was a complete shift of mindset: instead of something being true because Aristotle said it was, something became possible because of God's unlimited power. For those scholars with diminishing eyesight or problems with the eye, the invention of eyeglasses in Italy around 1280 must have been very welcome indeed.

As the fourteenth century began, crop rotation became standard farming practice in Europe. Depending on location and soil fertility, crops were rotated on either a two- or three-field basis. While farmers were seeing increased levels of productivity, sailors were making use of newly created nautical charts that featured wind directions and compass headings. Medical learning benefited greatly from the reintroduction of human dissection at the University of Bologna (earlier would-be physicians received their training via animal dissection and then extrapolated from animal anatomy to their human patients). Well-trained physicians were crucial with the increasingly routine use of gunpowder weapons such as cannon by 1324 and early handguns by 1326. Not all technology of this time was destructive. The mechanical clock, which brought the famous tick-tock sound to Europe, was created around 1330. Within a few years of its invention, municipalities competed with one another to have the best clock in their town squares. Milan was one of the first to boast such a technological marvel, in 1336. The invention of double-entry bookkeeping around 1340 did not influence most people in their everyday life, but it certainly allowed for easier commercial exchange among merchants and nations.

The fifteenth century was one of miniaturization and mechanization. With regard to the former, around 1430 the first spring-drive pocket watch went into production. While still individually crafted items and tremendously expensive, the ability to carry time with one made watches the must-have item of the well-to-do. By 1450, cannon powered by gunpowder had fully replaced the trebuchet as the field artillery piece of choice. Town and castle walls were redesigned to cope with the change. Walls were now sloped to deflect incoming projectiles and made of increasing thickness to absorb whatever was shot at them. Handguns, too, became a regular part of army equipment, and the age of mounted-knight warfare ended. One could learn of the latest battles and other news thanks to the inspired invention of Johannes Guttenberg, who, between 1452 and 1456, produced the world's first printed Bible on a press of his own design. Although not a big seller, the Guttenberg Bible proved the power of the printing press, and by 1460 Europe had presses in all major cities. The century ended with Christopher Columbus's famed voyage in 1492.

The sun set on the medieval period in the sixteenth century, which was both an end and a beginning. In 1512, the Catholic Church held its Fifth Lateran Council. One of the items on the agenda was calendar reform, because the Julian calendar, in use since 46 B.C.E., the time of Julius Caesar, was several days out of date, which had major implications for those of the Christian faith. The most crucial of these was the difficulty of determining the exact date of Easter. One of the mathematicians invited to participate in the endeavor to create a more accurate calendar was a Polish scholar named Nicholas Copernicus, who declined the invitation on the basis that astronomy needed to be better known and perhaps reformed prior to any alteration in the calendar. The new calendar, known as the Gregorian calendar, finally was instituted in 1578. However, nearly 30 years earlier, in 1543, the results of Copernicus's reevaluation of medieval astronomy culminated in the publication of his *On the Revolution of the Heavenly Spheres,* a book that is traditionally seen as initiating the Scientific Revolution of the seventeenth century. The following year, a French physician, Ambroise Paré, argued against treating gunpowder wounds with boiling oil. Instead, Paré advocated dressing the wound with cleansing agents, and soldiers all over Europe breathed easier.

From this brief historical outline, it should be readily apparent that medieval Europe accomplished much in the way of science and technology. While not all of the inventions (e.g., cannon and gunpowder) were for the benefit of humanity, many of the innovations created during this time in European history are still with us today.

CHRONOLOGY

793	Paper made in Baghdad.
830	"House of Wisdom" founded in Baghdad.
ca. 900	Baghdad has 30 public libraries.
	Introduction of wheeled plow, horseshoes, and horse collar into medieval farming.
900–1000	Introduction of paper into western Europe.
ca. 1000	Nearly the entire Greek corpus of medicine and science translated into Arabic and studied by Muslim scholars.
1000–1100	Peak of Romanesque church design.
1037	Avicenna, Arabic commentator on Aristotle, dies.
1066	Comet seen over England and taken as a bad omen.
	Norman conquest of England.
1086	Domesday Book in England records more than 6,000 waterwheels in the kingdom.
1095	Pope Urban II calls for the First Crusade to retake the Holy Land.
1098	Founding of Cistercian order of monks, who become important users of technology.
1100–1200	Peak of Gothic architecture in Catholic churches.
1140	Knowledge of magnets in western Europe.
1145	Legend of King Prester John established; for many explorers it becomes the impetus for their adventures.
1147	Possible first use of a trebuchet by European armies.
ca. 1150	Establishment of silk weaving in Italy.
1158	First European university founded at Bologna.
1165	Possible first use of a counterweight trebuchet.
ca. 1170	Flying buttresses used on the Gothic Notre Dame Cathedral.
1176–1209	Construction of London Bridge.
ca. 1180	Earliest mention of windmills in western Europe.
	First recorded mention of a compass in Europe.
	Advent of the sternpost rudder.

1185	First mention of a chimney in England.
1198	Death of Averroes, Arabic translator and commentator on Aristotle.
ca. 1200	Arabic numerals used in western Europe for the first time.
	Emergence of banks in Italy.
	Bulk of Aristotle's philosophy available in Latin translation.
	Scholasticism becomes the dominant form of scholarship in universities across Europe.
	Invention of buttons.
1200–1300	Holy Roman Empire (Germany) becomes the mining center of Europe.
1202	Zero introduced as a number in Europe for the first time
1209–1215	Incorporation of the University of Paris.
1210	Earliest restrictions placed on teaching Aristotle in universities.
1215	Magna Carta signed in England.
1220	Creation of armored helmet for knights known as "Great Helm."
ca. 1230	First depiction of wheelbarrow.
1249	University College, Oxford, founded.
ca. 1250	University of Bologna becomes center of medical studies in Europe.
1253	Death of the Franciscan friar Robert Grosseteste, who contributed much to the medieval study of optics and light.
1255	Lectures based on Aristotelian philosophy are made mandatory at the University of Paris.
1257	St. Thomas Aquinas starts teaching Aristotelian philosophy.
1265	St. Thomas Aquinas writes *Summa contra Gentiles*.
ca. 1267	Roger Bacon produces one of the earliest European accounts of gunpowder.
1274	Death of St. Thomas Aquinas, one of the earliest writer who attempted to reconcile Aristotle's philosophy with Catholic theology.

1275	Paper mills operating in Spain and Italy.
1277	Famous Condemnations of 1277 issued by the Bishop of Paris against much of Aristotelian philosophy.
ca. 1280	Eyeglasses invented in Italy.
1292	The Franciscan scholar Roger Bacon, who championed an experimental method to the study of nature, dies.
ca. 1298	Marco Polo returns from Asia and sets his experience to paper.
ca. 1300	Crop rotation established throughout Europe.
	Hammer-beam roof design becomes a common method of construction.
	First nautical charts appear.
	Human dissection practiced at the University of Bologna.
1315	Great Famine in Europe.
1317	Death of Salvino degli Armati, purported inventor of eyeglasses.
1324	First recorded use of gunpowder to propel projectiles in battle.
1326	First mention of handguns in medieval records.
ca. 1330	Likely date for creation of mechanical clock.
1336	Milan has a tower clock.
1337–1453	Hundred Years War between France and England.
1338	Earliest depiction of a sand clock.
ca. 1340	First appearance of double-entry bookkeeping.
1347–1352	Black Death ravages Europe, leaving 30 percent to 50 percent of the entire population dead.
ca. 1350	Northern European cog-style ships incorporate Mediterranean-style caravel ships to produce a carrack, the forerunners of galleons.
1360s	Italian armies experiment with handguns.
1370–1377	Construction of medieval Europe's longest stone bridge, in Italy, over the Adda River.
1410	Rediscovery of Claudius Ptolemy's *Geography.*

ca. **1430**	Spring-driven pocket watches appear.
ca. 1450	Gunpowder cannon replace trebuchets.
	Handguns form part of regular army complement.
1452–1456	Production of first printed Bible, by Johannes Guttenberg.
1460–1470	Printing press becomes more common in Europe.
1468	Death of Johannes Guttenberg, inventor of the printing press.
1475	Muzzle-loaded guns invented in Italy and the Holy Roman Empire.
1492	Christopher Columbus discovers the New World.
1499	Death of Marsilio Ficino, proponent of medicine that incorporated Platonic philosophy interpreted within a Christian framework that endeavored to drawn down stellar influences as curative agents.
1512–1517	Fifth Lateran Council sets out to reform the Julian calendar.
1517	Martin Luther's "95 Theses."
1537	Nicola Fontana invents gunner's quadrant.
1543	Nicholas Copernicus publishes *On the Revolution of the Heavenly Spheres*, the book seen as starting the Scientific Revolution of the seventeenth century.
	Death of Copernicus.
1544	The French physician Ambroise Paré argues against treating battle wounds with boiling oil.
1546	Paracelsus, advocate of a philosophy of nature based on chemical processes, dies.
1582	Gregorian calendar established by Pope Gregory XIII in papal decree *Inter gravissimas*.

I

EARNING A LIVING: AGRICULTURE AND MANUFACTURING

Despite the frequent focus in histories of the medieval period on people who did not have to work for a living but rather drew income from owning land, about 95 percent of Europeans had to support themselves and their families through varieties of labor. While not all work done in medieval Europe related to agriculture, farming and its associated jobs dominated the era's employment scene by accounting for the livelihood of about 90 percent of the population. Other types of work in medieval Europe included the production of beverages and other necessities of life, such as clothing.

This chapter opens with a brief description of the complex fabric of medieval society and the institutionalized system of expected duties known as feudalism. Social hierarchy enshrined by feudal obligations was also reflected in clothing, and thus the process of making fabric and clothing is examined next. Agriculture was the chief source of income within all strata of medieval society, and various aspects of it are presented in this chapter, including crop rotation and technologies of farming: plows, horse collars, and horseshoes. Medieval industry and society itself functioned through physical labor, and the theories and justification of labor are discussed here. Bread was the basis of the medieval diet, and making it often brought a wide segment of society into contact with the technologies of waterwheels and windmills. When stones were needed for building projects, iron for tools, and silver for coins, medieval Europe turned to mines and miners. Wine and beer were common beverages and were drunk at most meals; brewing them occurred both at home and at the

industrial level. Numbers, bankers, and bookkeeping kept mercantile
Europe in operation. While double-entry bookkeeping and the introduction of Arabic numerals meant little to peasants in the field, the ability to
keep track of expenses had an enormous effect on Europe in general.

FEUDALISM

Medieval Europe had a rigid social structure (especially in the countryside) that is known as feudalism. As the historian Jacques Le Goff describes
it, "the feudal concept of society rested on a sense of hierarchy which
expressed itself in vertical bonds held together by the oath of fidelity sworn
by inferiors to superiors" (Le Goff 1988, 90). Coming into full operation by
the eleventh century, feudalism appeared in most European countries.
At its most basic level, in a feudal system, a lord would grant the right to
use land to a social inferior, though still an aristocrat, who was called a
vassal. The land was known as a fief in this system. As an example, all land
in the kingdom belonged to the king, who, in the role of lord, would parcel
it out as fiefs to nobles, in the role of vassals. In exchange for the fief and
a promise of protection, the vassal contracted to provide certain days of
military service to the lord. This agreement was solemnized in a ceremony
called an investiture. Prior to the thirteenth century, the investiture consisted of an oral contract only, which reflects the importance of the spoken
word in the medieval world. After the thirteenth century, the investiture
was set on parchment. Those who could provide military service, usually
knights, were participants in the system, and those who could not fight
were not. Also, the vassal had to appear at the lord's request and pay hom-
age to the lord by participating in periodic ceremonies that demonstrated
the superior position of the lord in the relationship. Each participant in the
feudal structure owed duties and in turn was owed duties.

Vassals could also be lords. Vassals to the king who controlled great
amounts of land could, in turn, divide up that land and offer it to nobility of lesser standing than themselves, and the whole process would be
replayed further down the social hierarchy. Even kings could be vassals.
English monarchs, to cite a famous example, were at the top of the feudal
system in England but also ruled land in France, which made them vassals
of the French king. Dissatisfaction with this situation on the part of English
kings was one of the chief causes of the famed Hundred Years War.

Feudalism began to wane around the fourteenth century, when military service to the lord granted by the vassal was converted to monetary
payments. Warfare itself changed with the decline of the mounted knight
as the chief combat force. What is more, the rise of towns and commerce
provided new means of livelihood and generated forms of wealth that
were not tied to the land. Kings, not wishing to be beholden to their nobles for military service as outlined in the feudal system, turned to hiring
mercenary armies for cash instead of seeking their service in exchange for
land (Le Goff 1988, 90–95; Reynolds 1994, 17–22).

The place of peasants, who in some estimates made up around 98 percent of medieval society, in feudalism is not exactly clear. Some scholars argue that peasants were part of the feudal system, while others argue that feudalism applied only to the warrior class of medieval society. While they occupied the bottom of the social structure and worked land belonging to a feudal lord, peasants were tied to the land they worked and were not truly free to participate in the other arenas of feudalism. Certainly a peasant could not be a lord to a vassal because peasants did not have land to grant as fiefs and no one ranked beneath them in the society.

This vertical structure is reflected in other aspects of medieval society, too, as the pioneering work of the historian Marc Bloch has demonstrated. Most prominently, it is seen in religion. God sat at the top of Christianity, followed by angels, the pope, then bishops, priests, and, last, everyday Christians. Church officials could also be lords in the feudal system, thus making them part of two hierarchies. The thing to remember about medieval society is that social standing was granted at birth; moving up to a higher position was practically impossible. There were three broad categories: those who fought (kings, knights, and nobles), those who prayed (pope, cardinals, bishops, and priests), and those who worked (everyone else). In modern society, social position comes from the possession of financial means and the ability to purchase certain items, such as foreign cars, big homes, and designer clothes. Those who can do so are said to be of the same social rank. In medieval Europe, title was everything. Even if merchants made a fortune in the spice trade, they would always have less standing than even the poorest noble because that was the way society worked. Upward mobility was practically unheard of, and the lack of possible advancement into avenues of society reserved for nobles or those with a title, such as a role in government, would prove one of the major sources of discontent in the coming decades and centuries. While there were a few shadings to the description I have given (e.g., some peasants owned land and had different rights granted by their lords), generally they are the exception that prove the rule.

It must be noted that the term "feudalism" was invented in the seventeenth century and was not used in the medieval period. This fact has led some historians to deny the existence of feudalism in the medieval period because broad categorizations of feudalism do not correspond exactly with what is seen in all situations in medieval Europe. Susan Reynolds has most recently argued this view (Reynolds 1994). Other scholars counter that, even if the term did not exist, the concept did, and they view the debate as a matter of semantics.

CLOTHES AND CLOTH MAKING

Most everyday medieval fabrics were wool or linen made from the fleece of sheep and flax plants respectively. The warmth of wool made it a natural choice for outerwear, while linen cloth was used to make underwear. (When they wore out, linen clothes were recycled in the paper-making process.) The quality of wool depended on the part of the sheep from which the fleece came: fleece from the shoulders was best, and fleece from the belly was least desirable. To make woolen cloth, the first step, after shearing the sheep, was to comb the fleece with heated woolcombs that had been coated in a mixture of oil and animal fat. This process produced uniform fibers ready for spinning. Next, the strands of wool were spun to form thread. All this was done entirely by hand. The threads were wrapped around a small rod called a drop spindle and, later, around a simple spinning wheel. As soon as enough thread had been collected, the process of weaving began. On early medieval looms, which were vertical looms, the individual threads were kept taut by weights attached to their ends. Cross-threads were woven between the vertical threads one thread at a time. The introduction of the horizontal loom sped up the process by replacing the weights with rollers that maintained tension on the vertical threads (Singman 1999, 35–36).

More luxurious material used by the wealthier members of society included imported silk in the form of clothing and bed sheets. Owning silk items represented power and authority. Until around 1150, when the first European silk-weaving capabilities were established in Sicily, medieval silk came from the Byzantine Empire and from regions under Islamic influence. Silk was a major component of the Mediterranean economy as merchants and traders brought the sought-after fabric to markets throughout Europe. Weaving silk was much more complicated than weaving wool or linen. Silk was rarely made into a plain fabric of a single color; very often, it was decorated with birds, human figures, and geometrical patterns. The skill needed to weave these images into the finished piece of fabric, in addition to the transportation cost, ensured that prices remained high. With the founding of European silk manufacturing in the thirteenth and fourteenth centuries, silk did begin to trickle down the social hierarchy—but not very far.

Once a piece of woven cloth was completed, it had to be finished. Woolen cloth needed to be cleansed and the fibers tightened up before it could be made into a piece of clothing. The best way to remove the lanolin (or natural grease) from woolen cloth to make it smooth and soft was to soak it in a solution of ammonia. Ammonia occurs in great amounts in urine. People known as fullers cleaned the wool by immersing it in barrels of stale urine and mixed it by stomping on the cloth while it sat in the urine. After the woolen cloth had dried, the next task was dyeing. Often wool thread was dyed prior to being woven, but not always. Medieval dyes were all natural, made from substances such as leaves, plant matter, and crushed bodies of insects, and they produced somewhat dull colors. Cloth or thread was soaked in a mixture of dye and hot water. The desired intensity of the color determined the length of the process. Most everyday people wore linen in its natural brownish color. Wearing white linen, however, was a status symbol, and well-off people paid a great amount to have their linen bleached. For example, King Edward II of England employed two laundresses to ensure a supply of bright white linen for the royal household. Bleaching was done either by leaving the cloth outside in the sun or by soaking it in a solution of lye. Medieval lye was a mixture of animal waste and human urine combined with limestone and wood ash (Crowfoot et al. 2001, 19–20, 81).

Not all medieval fabrics were woven. Archaeologists have unearthed knitted garments dating from the fourteenth century. Other nonwoven woolen fabrics included felt. Felt was made by the "matting together of fibers, with moisture and heat under pressure, to produce a solid non-woven fabric" (Crowfoot et al. 2001, 75). The result was water- and windproof cloth. Common uses for felt among the upper classes were for making hats (particularly in the fourteenth century), boots, and outerwear; it was also used in the home for carpets and tapestries.

The clothes worn by early medieval people were generally loose and pulled over the head. There was not the level of tailored clothing that is common in the modern world. Only around the fourteenth century, with the introduction of clothes that opened in the front and closed with buttons, did structured and tailored pieces become more common, although mostly among the upper strata of society. Buttons first appeared in Italy in the thirteenth century and then in the rest of Europe by the fourteenth. Initially, buttons were not the practical fastening devices they are today but rather expensive pieces of jewelry. Inexpensive buttons were made out of copper, glass, or balled-up bits of cloth. While much medieval clothing was tunic-style and was worn gathered around the waist with belts made of either leather or cloth, some sewing was required. Sewing thread used in medieval clothing was made from linen. On more expensive pieces of clothing, the thread was silk, but this would not have been used on more than 90 percent of medieval clothing. Cotton thread seems not to have been used in any great quantities. Iron needles were used to pull the thread and make the seams and, later, buttonholes. Making needles was the task of some of the lowest members of society. Great dexterity and small hands were needed to turn thin iron wire into sewing needles.

CROP SYSTEM

Between approximately 600 and 1000, the population of Europe increased by nearly 38 percent. This drastic increase provided the impetus for new and more productive methods of agriculture. In the northern European countries of England, France, and Spain, the population expansion was very significant. However, the increase in population led to a corresponding demand for grain, which forced farming into new areas that were not well suited to the strains that the process placed on the land. For example, following 1066, many of England's marshes, moors, and forests were drained and cleared to make way for more farms and

living space (McClellan and Dorn 1999, 177). Much of the reclaimed countryside could support crops for only a few seasons before the nutrient
content of the soil was exhausted. By expanding into unprofitable parcels
of lands, medieval Europeans left themselves susceptible to a series of
poor harvests, resulting in widespread famine, such as the Great Famine
of 1315–1317, the episode many scholars see as preparing Europe for the
disaster of the Black Death in 1347.

One response to the demands of population growth was a new method
of farming that involved variation in planting. Under a system of crop
rotation, different crops are planted on specific sections of farmland. The
goal of rotation is to prolong the fertility of the land by not exhausting it
through consistent planting of the same crop on the same parcel of land
year after year. A simple form is a two-crop rotation where one field is
planted and the second is allowed to sit fallow (without plants). Animals
are permitted to graze on the fallow land and fertilize it with their waste.
Romans, for example, used a two-field crop rotation system.

More complex systems were based on a three-field rotation, which
appeared in Europe during the eighth century. In the ideal version of this
scheme, a parcel of land is separated into three equal sections. One field
is sown with a winter crop (wheat or rye), one field is planted with a
spring crop (barley or oats), and the third field is left alone for animals
to live on and fertilize. The fields then rotate every year or every other
year, depending on the particular farm. There are two advantages to this
system. First, only one-third of the land is unused, instead of half as in
the older two-field system. Second, scholars estimate that rotating crops
increased farm yields by at least 30 percent and by perhaps as much as
50 percent (Rösener 1992, 43, 117). However, modern notions of the three-
field crop rotation system (where land is divided into two or three large
parcels) might not have existed as a true model in the medieval period. In
the example of Flanders, farmers divided their land into smaller sections
(often these divisions were imposed upon the land by natural obstacles
or by human-made obstacles such as roads), which were planted with
various crops (Astill and Langdon 1997, 74). What is more, with regard
to the division of land surrounding the manor house, at least in England, it is important to remember that there were not three large sections
of land. Several parcels of land were divided among the peasants, and
within these parcels the rotation of crops took place. Winter crops might
be sowed next to spring crops, or a fallow field might be fit between the
two. Peasants might even have their small sections of land spread across
an entire field.

Whatever form it took, most of Europe adopted some form of crop
rotation. In France, for example, the three-field rotation dominated farming
practice. A two-field crop rotation system was introduced in Sweden around
the year 1000 and was the chief form of planting by 1300. That crop rotation
differed from one area to another may be explained by the variations of soil

and climate. Take the case of Cistercian monasteries: monks in the north-ern regions of France employed a three-field rotation, while their brethren nearer the Mediterranean used a two-field system with bountiful results. Animals could also increase the yield of farms through the deposition of manure on fields that lay fallow. Cistercian monasteries found that as the number of animals owned by the brethren increased, so too did the quantity of crops that grew on their farms (Berman 1986, 91, 94).

Many historians view the innovation of rotation crops as tremendously important in the creation of the medieval agricultural revolution. Recently, this assertion has been questioned by the scholar George Comet, who argues that, for a specific section of land, rotating crops did not increase the productivity of that land. The benefit of the system, as Comet sees it, is that planting winter and spring crops lessened the risk of crop failure by ensuring that a bad spring, for example, did not wipe out the entire year's yield (Astill and Langdon 1997, 29). Nevertheless, the historical consensus points to crop rotation as something quite revolutionary.

FARM TECHNOLOGIES: PLOW, HORSE COLLAR, AND HORSESHOES

The plow was arguably the most important medieval farming tool, because without the ability to effectively place the seed in the ground the entire farming enterprise was moot. The earliest method of getting the seed into the ground involved nothing more than hard labor with a shovel. Even the poorest farmer would have been able to afford some hand tools. With a great deal of work, hand tools could cultivate a suffi-cient amount of farmland to ensure a family's survival. Some of the basic farm tools included axes, shovels, pickaxes, and hoes. These tools were all made of iron except for the pitchfork, which seems to have been made of wood (Astill and Grant 1988, 95).

Early medieval plows were known as scratch plows or "ards." These devices may or may not have had wheels; the evidence is too sketchy for us to say with any certainty. An ard, which was basically a sharpened wooden post, broke the ground for planting by scratching the surface of the soil, but it did not turn over the soil. While the ard was pulled forward by either animal or human power, a farmer had to press constantly down on the ard to ensure that the sharp edge maintained contact with earth. In areas with soft soil, the ard worked quite well. However, in northern regions, with a more firm soil, it did not prove as effective.

The wheeled plow was a major innovation because it allowed more land to be quickly cultivated, particularly sections that had been too hard to break using earlier methods. The plow consisted of several parts. First, a cutting knife sliced into the top sod. Second, a plowshare cut the sod hori-zontally. Third, a moldboard turned the cut sod to one side, revealing the soil into which the seed was sown. (Think of a snowplow pushing snow off to the side as it clears streets in winter.) Wheeled plows allowed the farmer to lock in the level of the plow blade in relation to the wheels and thus ensure a constant depth for planting. The use of this type of plow and the density of the land on which it was used required a large number of animals. Because the teams of oxen or horses, or combinations of the two, were difficult to turn frequently, the shape of farm fields changed from square to more rectangular. The wheeled plow also spurred the development of collective farming enterprises, because very few farmers could have owned a plow and teams of animals by themselves. The wheeled plow, in combination with the horse collar, is viewed by many historians as the key factor in any description of a medieval agricultural revolution. The wheeled plow is noted in thirteenth-century literature, but it was likely well established before this date. Efforts to pin down exactly when this type of plow replaced less advanced technology are made difficult by the unchanging terminology of the period, which tended not to distinguish one type of plow from another (Astill and Langdon 1997, 59).

Through the tenth and eleventh centuries, the horse assumed great importance for farming. However, there was tremendous diversity in the effectiveness of the horse that may be traced to the harness or collar used. Early historians of farming technology celebrated the padded horse collar as a major invention, one that, on its own, redefined the practice of medieval farming. While this praise has become less effusive in recent years, the collar remains an important piece of medieval technology and one that rightly merits attention.

Farmers of antiquity relied on a team of two oxen to pull a plow and created a yoke so that the animals could pull either the plow or a cart. Oxen were among the first animals used as a power source; their use dates to between 4000 and 3000 B.C.E. Horses, on the other hand, were first harnessed some 2,000 years later. The equipment required to harness oxen was relatively simple: a wooden yoke was placed around the animal's neck. Here anatomy plays a role in technological development. The spine of an ox is quite bony and area between the protrusions was a natural location for a yoke. Oxen were then used as side-by-side pairs with the plow attached by a shaft between them (Langdon 1986, 5–7). However, horses have greater endurance than oxen and can work on average two hours longer per day. As the historian John Langdon points out, the replacement of oxen by horses in farming was a great technological innovation, even though one animal simply replaced another.

This increased power could be utilized for farming if horses had the right harness. Horses are obviously not oxen, and a rigid yoke that worked on an ox would not work with the same effectiveness on a horse. Indeed, the ox yoke would constrict the longer neck of a horse. Horses have smooth backs and an upright posture that was better exploited by some variety of collar or early attempts at a throat-and-girth-harness. The drawback with the throat-and-girth-harness was that the straps often

Wheeled plow. The Pierpoint Morgan Library, New York. MS M.399, f. 10v.

shifted position as the horse moved and could potentially choke the animal. The rigid collar solved this problem; it was placed over the horse's head and rested on its shoulders. This permitted unobstructed breathing and placed the weight of the plow or wagon where the horse could best support it (Langdon 1986, 9). The new collar was introduced in Europe around the eighth century, and the earliest depiction of this innovation is seen in the famed Bayeux Tapestry, created in the eleventh century. There was a steady increase in the number of horses used for farming, and by the fifteenth century horses accounted for around 30 percent of animals used to pull plows and other heavy hauling tasks.

A horse pulling a plow required secure traction on a variety of surfaces. Footing was a major issue in farming and indeed in all areas of horse use. The practice of shoeing a horse had been widespread during the period of the Roman Empire. Most historians believe that this technology had been lost by the start of the medieval period and that it was not recovered until the late years of the medieval era. However, some scholars argue that no evidence exists to defend this lost-technology thesis and that the practice of shoeing horses continued throughout the entire medieval period.

The horseshoe is best seen as hoof protection for the horse, a device to prevent undue wear and tear on the hooves as the horse works on rocky and uneven land. There does seem to have been a proliferation of horse-shoe use in the tenth and eleventh centuries. This has been attributed to the rise of hard-packed medieval roads in areas where transportation re-lied greatly on horsepower. Here, commerce and profit motivated peo-ple to care for the hooves of horses that were expected to carry goods to markets, often over great distances. Horses fitted with shoes could also be counted on to carry and pull far greater loads than would have been possible with bare hooves. What is more, shoed horses were able to point the front edge of their hooves and dig them into the ground to aid in the forward motion (Leighton 1972, 107–108).

For the cultivation of soft farmlands, the horseshoe perhaps was not as important as past scholars have believed. Nonetheless, in harder European climates, horseshoes were required if the land was to be fully cultivated. This was certainly true in the climatic varieties of northern Europe. Pro-longed periods of cold and wet winters tended to soften and weaken the hooves of horses, which then became lame. It was not unusual for farmers to shoe only the front hooves of their horses. From this practice it seems most likely that the benefit sought by shoeing plow horses was increased traction. By the fourteenth century, the technology of shoeing had become so entrenched that even oxen and donkeys were shoed to decrease the wear on their hooves and to improve their footing while pulling plows and carts (Langdon 1986, 10).

The use of horseshoes has been cited by historians as a marker of the proliferation of iron in medieval towns and villages. Between 1000 and 1300, the weight of the horseshoe, in England for example, was around a quarter pound, and the shoe was attached by six nails. By the fourteenth century, the weight of the horseshoe had increased to nearly half a pound, and the shoe was fixed by up to eight nails. The gain in the size of the shoe and the number of nails required to hold it on the horses' hoof suggests that iron production and the prevalence of smiths in various localities were both increasing.

LABORERS

Medieval Europe functioned on the backs of laborers. While planners, architects, knights, and other, more visible members of society get all the credit and acclaim for developments in the science and technology of the era, few of these would have occurred without people to do the dirty and unpleasant work. Labor was institutionalized in medieval Europe;

it was part of the social structure. What is more, religious salvation often required labor for the good of one's soul.

In the feudal system of land ownership that dominated the early medieval period, the majority of farmers were "dependent peasants," to borrow the terminology of the historian Werner Rösener. Such people were required to perform a great deal of labor for their lord, who owned the land upon which the peasants worked both to provide for their own families and to fulfill feudal obligations. Although the peasants were required to labor for the lord, they were expected to provide the tools and animals with which they did so. As Rösener notes, the expected duties of peasants were extreme. The lord's peasant farmers had to plow the land, sow the seed, harvest the crop, grind the grain, and bake bread. In their spare time, they were to build fences, chop firewood, drive carts, and do any other chores required of them. However, some innovations did ease the workload carried by peasants, at least a little. The three-field system allowed the labor of farming to be more evenly spread out over the course of the year, rather than being condensed into a few months. The introduction of animal power into the process of plowing fields and the increasing prevalence of this practice greatly reduced the manual labor involved in sowing the crop. The demands placed on the peasant farmers by their lords were among the chief causes of revolt in the medieval period. When peasants believed that the labor requirements placed upon them were too severe, resistance was certain. On the other side of the fence, lords were known to complain about the lackluster efforts made by their peasants when performing their laborious tasks (Rösener 1992, 53, 119, 243).

By the thirteenth century, many of these labor-based duties had been replaced by monetary payment to the lord in lieu of physical work. The process was fully under way in the fourteenth century. Landlords in England had converted many labor requirements from their tenants into cash payments. In part this alteration came from the dramatic increase in the cost of labor as a result of the Black Death. Working for money was frequently the only option for the poorest of peasants, those who did not own any land, and the second sons of wealthier peasants had to seek other employment than farming. The likelihood of finding paid work as a laborer increased throughout the medieval period and was quite good in villages. There were two main types of labor services: long term and day labor. Increased demand for labor at certain times of the year, such as during harvest and seeding, attracted many outsiders who were desperate for work to communities (Schofield 2003, 32, 125–126).

Manual work was also considered good for one's soul. Members of monastic communities performed manual labor as part of their devotional activities. To work with one's hands for the glory of God was a form of worship and piety. Thus, labor was also a necessary aspect of monastery life. While prayer was the pinnacle of a monk's life, the realities of everyday life—making repairs to the monastery, performing chores, gathering foodstuffs—meant that monks did not live by prayer alone. Church fathers like St. Augustine endorsed labor because they believed that, through physical engagement with God's creation, humanity may come to know the providential plan encoded within the world around them. Later Catholic scholars such as St. Thomas Aquinas suggested that labor warded off the sin of sloth and taught humility (Ovitt 1987, 199–200). While this may have been the idealized depiction of labor in medieval Europe, the realities of toiling in a field, carrying mortar, or setting stone likely obscured this grand interpretation.

WATERWHEELS AND WINDMILLS

Grinding wheat into flour was an everyday task in medieval Europe. For peasants, the job involved placing wheat between two round stones (millstones) and turning the top one to produce flour. This is where the expression "daily grind" originates; it could take all day to get enough flour to make bread to feed a family. And, since bread was the major component of the medieval diet, the task was done every day. The other alternative was to use a mill powered by a waterwheel or a windmill. However, these technologies were expensive and beyond the means of almost all individuals. Traditional accounts of mill (water and wind) ownership place them within the hands of the great noble landlords. While it is certainly true that lords owned a majority of mills and charged their peasants extortion-level fees to use them, other groups in medieval Europe, such as peasant collectives and monasteries, also operated water and windmills.

The Roman Empire made extensive use of the waterwheel to grind large quantities of grain into flour. Evidence of Roman waterwheels is found as far north as Hadrian's Wall in England (just south of the English-Scottish border). This technology reached its zenith with the flour-producing factory located at Barbegal, in southern France, in the fourth century. The mill at Barbegal could yield 31 U.S. tons of flour every day. England in the eleventh century had many water wheels. The Domesday survey, initiated by William the Conqueror in 1086, counted just over 6,000 waterwheels in England. Over a third of all manor houses owned a wheel, which served the lord and his peasants (Holt 1988, 2; Langdon 2004).

The two basic types of waterwheels are the undershot and the overshot. Often it is unclear to which types of wheel medieval authors are referring in their writings. The undershot is a vertical wheel that sits in a body of moving water (most often a river), using the power of running water to make it turn. The overshot relies on a supply of water from above and catches the water to make it move. Historians believe that the undershot variety of wheel was prevalent in medieval Europe. Both types require a system of gears to turn the vertical motion of the wheel into horizontal motion in order to work a grinding stone. What is more, gears could also speed up slow water or slow fast-running rivers. Whatever speed was needed to effectively run the mill could be obtained. The axle attached to the center of the waterwheel sat on iron bearings that allowed for smooth rotation and cut down on the wear on the wooden axles. Many of the waterwheels used in medieval Ireland were different from those in other European nations. Irish wheels were horizontal, with the grinding stone placed directly above. This configuration avoided the gearing required in vertical models (Hill 1984, 156–157; Holt 1988, 4).

For the first 800 years of their use, waterwheels were used specifically to grind flour. Some enterprising millers experimented with double waterwheels, each powering a millstone. However, this innovation seems not to have been very widespread. Part of the problem with this design was that the differential (a gear that allows two shafts to rotate at different speeds; it is used on modern vehicles so that the rear wheels can turn corners at different speeds) was unknown to the medieval craftsman; as a result, any problem with one wheel meant that the other would also be affected, thus negating any advantage of having two wheels.

The enormous power output provided by waterwheels was used in multiple ways. Acquiring the ability to direct the rotational motion delivered by the waterwheel was a major innovation. The same vertical motion of the waterwheel that was transmitted into horizontal motion to operate a millstone could also operate a crank to power a saw to cut more wood to make more wheels. One major tool run by the waterwheel was the cam, which worked a trip hammer to soften cloth or smash ore harder and faster than could be done by hand.

For areas that did not have easy access to running water but still required flour for bread, the windmill was the solution. Windmills were likely unknown to earlier civilizations. The windmill is a unique piece of medieval technology because it almost certainly is a medieval invention rather than a modification or a rediscovery of an earlier technology, be it Roman or Chinese. References from ninth-century Baghdad describe a small windmill that was used to run a fountain. Unsubstantiated reports from tenth-century Afghanistan speak of a much larger windmill. More reliable accounts from Syria in the thirteenth century describe a modern notion of a windmill built specifically for grinding grains. Western Europe used windmills from about the twelfth century.

Archaeological evidence reveals that medieval windmills existed along the southern English and northern French coasts, as well as in some areas of Flanders (near modern Belgium). In the thirteenth century, the technology of windmills reached Holland, which embraced it readily and constructed a great number of the structures. Windmills had at least one advantage over their water-run counterparts: they could continue to operate in the winter months, when watermills were literally frozen in their tracks. From around 1180, many contemporary documents discuss windmills. Consider the assessment provided by the historian Jean Gimpel: "There were so

Medieval landscape with windmill. The Pierpont
Morgan Library, New York. MS 399, f. 8v.

many windmills, bringing in such high profits, that Pope Celestine III
(1191–1198) imposed a tax on them" (Gimpel 1976, 25).

Medieval windmills did not operate like their modern counterparts.
Whereas contemporary windmills function with blades on a pivoting
base that rotates to face the wind, medieval versions were more complicated. As Donald Hill, a historian of engineering, explains, "Medieval
windmills were post-mills, in which the whole structure is rotated in
order to face the sails into the wind. The tower mill, having a rotating
cap with a fixed structure, may have been introduced towards the end
of the fourteenth century but did not come into general use for about
another two centuries" (Hill 1984, 174). As in waterwheel-powered
mills, the spinning of the windmill blades powered a milling stone that
was used to grind wheat into flour. Despite their similar purpose within
the medieval economy, it appears that waterwheel mills and windmills
complemented each other, rather than competed (Holt 1988, 34, 36).

MINING

Much early medieval mining was the surface or open-pit variety and required few special tools, although sometimes miners excavated shallow caves. Iron, the staple of medieval industry and technology, was most readily available in iron ore, rather than in veins under the earth. While not always visible, iron in the form of clips, ties, and rods helped hold together the stonework in the largest medieval buildings. Historians suggest that the earliest iron used in antiquity and the early medieval period came not from the earth but from meteorites, which have a high iron content. Another source of iron was bog ore. Finding these raw metals (which is a type of mining) has been described as one of the worst jobs in history because it consists mostly of poking the marsh ground with a stick until it hits metal. The poorest members of society did this task as a way to earn a very modest living.

Stone had been taken from quarries since antiquity and was a popular building material for medieval cathedrals and castles. (Like stone, coal too was frequently found in surface pits.) These products were relatively easy to obtain and required little in the way of excavation. Obtaining stone was the most important of all mining activities in medieval Europe. The European countryside was covered with many open-pit stone quarries; France boasted the most of them. Indeed, stone was one of France's chief exports to England (Hunt and Murray 1999, 45). Often, stone was broken with little more than an iron hammer. In a more advanced technique, the rocks were heated, and wedges were then forced into the expanding cracks to fracture the stone.

Smelting metals (copper was the first) began around 4300 B.C.E. in modern Turkey and Iran. By adding tin to copper (both are easily melted), metalworkers of antiquity produced bronze, an alloy that was considerably harder and more durable than either copper or tin alone. Bronze was ideal for church bells, as well as for swords and early armor. In later medieval times, bronze became the material of choice for cannon construction. The demand for bronze created a greater demand for tin, which became the subject of searches and mining endeavors (Gregory 1980, 63–64).

Mining in the modern sense of tunneling underground focused on recovering precious metals. The most frequently sought was silver for coins. Major silver strikes took place within modern Germany and Austria. The mineral rights belonged to the monarch or prince who ruled over the land in which the mine was built. These superseded even the property rights of the landowner, whose claim was only for the surface elements of the land and covered nothing underneath. Miners, too, benefited from privileges under the law that were not available to other members of society. Special law courts were established to hear grievances against miners who would often take whatever timber they needed and divert streams to meet their own needs. However, these courts were set up to give miners the benefit of the doubt, and compensation was rarely awarded.

During the thirteenth and fourteenth centuries, Europe's demand for metals, particularly gold and silver for coinage, sparked a concerted effort to find new supplies. Lengthy veins of silver in the Holy Roman Empire established modern Germany as the mining capital of medieval Europe. Places like Bohemia became among the wealthiest of all territories. As shallow supplies ran low, mines got progressively deeper. The excavation was still done entirely with hand tools. Only in the sixteenth century did miners first use gunpowder as a tool to blast solid rock. The increased complexity of mining meant that no single person could finance the operation. Securing investors and selling stocks became the common method of bankrolling a mine. Early underground mines were quite shallow by modern standards. Shafts did not extend below the level of ground water. As a result, drainage was not a problem. However, as the supply of shallow ore became exhausted, deeper exploration was required. Then problems of drainage came to the fore of mining. One of the first solutions to the problem was to dig a drainage shaft that effectively lowered the level of ground water in the mine. Other technological solutions to this problem included an Archimedean screw and a waterwheel. The waterwheel worked as follows: a large wooden wheel, powered by a worker walking within the wheel (think hamster wheel), rotated buckets or absorbent rags through the standing water and lifted them up to a reservoir on another level of the mine. Through a series of wheels and reservoirs, water could be lifted to the surface of a mine from great depths. The Archimedean screw, invented by Archimedes, removed water from mines by means of a large rotating screw (think auger) that moved water in the direction of rotation. With a collection of screws (each could move water only about six feet), water could be extracted from mines (Gregory 1980, 165–166).

MAKING DRINKS: WINE AND BEER

Winemaking and the cultivation of vineyards was well established in the ancient world; indeed, wine was a staple of Roman diets. Following the breakup of the Roman Empire, knowledge about winemaking survived in medieval Europe. However, the mechanism of that survival is the subject of scholarly debate. On the one side are historians who suggest that the importance of wine in the Catholic Church, as the symbol of Christ's blood and a key component of communion, ensured that information about wine and the technology needed to make it was guarded by the Church. Other scholars dismiss ideas of ecclesiastical altruism and point to the self-interest of wine merchants who saw wine as a commodity that was too valuable to simply be allowed to vanish from the medieval marketplace. Adding weight to this argument is the fact that Germanic tribes did not destroy Roman vineyards during the invasions that would ruin the rest of the empire. Thus, it is claimed that a thriving wine industry survived relatively unscathed without the Church's protection.

The care of vineyards and the harvesting of the grapes were tasks done by peasants. Whereas, in the region around the Mediterranean, almost any type of grape could be successfully grown, in the cooler climate of northern Europe only specific grapes flourished. Prior to the 1200s, most northern vineyards produced white varieties, which were better adapted to the climate. Over time, the cultivation of red grapes became a viable enterprise even in the north. Once the grapes were harvested, they needed to be pressed. The considerable cost of a medieval winepress suggests that only nobles or monasteries could afford one. However, a less expensive method involved the crushing of grapes under foot. For white wine, a press was necessary so that the color of the wine was not altered by prolonged contact with the skin of the grape. Such a concern did not apply to red wine, but a press was still needed even if the initial crushing of the grapes had been done underfoot, because the use of a press for the final stages squeezed the last drop of juice from the skins. Once the juice was extracted, it was poured into wooden barrels to ferment (Unwin 1991).

Medieval Europeans drank wine not only for pleasure and to quench thirst. Wine was thought to possess certain medicinal properties. Ancient physicians such as Galen, whose medical writings informed the teaching of medicine in medieval Europe, prescribed wine as a cleanser of wounds and to bring down fevers. Medieval scholars also suggested also that wine served as a satisfactory antiseptic.

Beer has been made and drunk since the earliest civilizations in Mesopotamia. Beer was a major component of the medieval diet; it was drunk at most meals, including breakfast. Basic brewing of beer is done in four steps, with the fifth step being packaging and delivery. First, a brewer malts the grain and then grinds it. This process involves germinating the grain in water, then halting the growth and drying the grain. Second, the ground malt is set in hot water to form a mash. The mixing of water and mash produces a liquid called wort. Third, the wort is then collected and boiled in the presence of additives, the most common of which in the medieval period was hops. Fourth, after being cooled, the wort is fermented with yeast.

Medieval beer production took place mostly at the household level. This often meant that the women of the house made beer as part of their everyday chores. Of these operations, the largest were those carried out in monasteries during the birth of the monastic movement around the eighth century C.E. Since monasteries produced surplus grain, they had ready access to the prime material of brewing. Some medieval religious writers believed that beer possessed medicinal properties. For example, Hildegard of Bingen wrote that beer consumption might cure lameness.

In the late medieval era, town brewers became more numerous, though the production of personal beer supplies within households continued. Nonetheless, with no end in sight to the trend of urbanization and the continuing popularity of beer drinking, urban brewers played an increasingly large role in the making of beer. By 1300, it was possible to make a living as a brewer. The large quantities of beer produced by brewers in urban settings required a greater initial investment in equipment. This was particularly true of the purchase of a copper kettle. Copper had advantages over the traditional materials of wood or pottery. The heat conductivity of copper meant that less liquid was lost during the wort-boiling stage (this meant more beer at the end), and less heat was required to keep the wort mixture boiling (this meant lower fuel costs). The large brewing kettles of the type considered here could hold somewhere around 250 gallons in the 1200s and more than 1,000 gallons by the 1400s. Once produced, beer traveled to markets all over Europe. By 1200, beer exports had become a chief source of income for several German towns. The method of transportation, however, determined the final selling price. Shipping beer over land added up to 70 percent to the cost of the beer because of the enormous expenses of carriage. Conversely, beer transported on waterways was substantially less expensive. This fact explains why port cities came to be synonymous with the beer trade, and, indeed, all trade (Unger 2004).

NUMBERS, BOOKKEEPING, AND BANKING

What to the modern eye looks like numbers (1, 3, 5, and the like) is of Arabic origin and did not exist in Europe until quite recently, historically speaking. Not surprisingly, western European scholars had continued to employ Roman numerals for centuries after the fall of the Roman Empire. Christian thinkers in Spain, which had a large Muslim population, first noticed Arabic numerals (which actually originated with Hindu scholars). Arabic numerals were used by Muslims in the 800s and appear in European manuscripts from around 976. In 1202, Leonardo of Pisa composed *Liber Abaci*, in which he outlined the advantages of Arabic numbers by comparing them to their Roman equivalents in adjacent columns. By 1240, Italian merchants had embraced Arabic numerals, and they were soon followed by the rest of Europe. Despite the enormous advantage of the Arabic system (try dividing XXVIII by V), many in medieval Europe resisted what they viewed as a heathen number system, and it was not until 1514 that the last mathematical treatise featuring exclusive use of Roman numerals was printed (Crosby 1997, 112–114). As the historian Alfred W. Crosby has noted, the reluctance to fully replace Roman numerals with Arabic led to kind of transitional phase where all sorts of hybrid number systems appeared: IVOII (for 1502), MCCCC94 (for 1494), and MCCCC4XVII (for 1447).

Even with the acceptance of Arabic numbers, the notion of zero was not easily understood. The term "zero" comes from the Arabic word *sifr* (the empty) through its Latin transliteration as *zefirum* to its Italian translation as *zefro* before acquiring its modern recognizable form. Zero was a difficult concept for medieval thinkers because it was both a number and nothing.

By itself, a zero signifies nonexistence, but when it is placed beside another number it has the ability to magnify that number by ten. Although initially rejected, the numeral zero would become an indispensable tool for mathematicians and bookkeepers (Flegg 1989, 126–127).

The oldest form of medieval bookkeeping arose because lords and other employers needed to keep tabs on their workers, in addition to monitoring the expenses of the manor. Also, local officials of the Catholic Church were required to maintain accurate accounts of tithes, taxes, and other monetary gifts. While many of these records were set down on parchment, this was not always the case. In pre-conquest England, virtually all business transactions were communicated orally and trusted to memory. Throughout the medieval period, most people's everyday business transactions consisted of oral contracts.

Early medieval merchants did not keep very detailed account books because the number of goods was limited. Beginning in the 900s, the nature and volume of trade increased. This change necessitated a more complex system of recordkeeping. Partnerships between merchants were a regular occurrence, with each individual ensuring fair treatment by noting the details of the arrangement with a new level of precision. Careful records were important because it was not uncommon for the trading arrangement to outlive the original partners. Paper lasts longer than human memory. Whereas merchants used to travel the seas themselves in search of riches, by the later medieval period they employed agents to do the legwork while they stayed home and kept track of transactions.

By the middle of the fourteenth century, Arabic numerals started to appear in ledgers and merchant account books. Prior to this time, account books tended to be narratives of transactions and debts, while afterward ledgers took on their modern appearance of columns of numbers. Double-entry bookkeeping first arose around 1340, and possibly earlier, within the circle of Italian merchants. This system recorded transactions in two columns: expenses on one side of the page and sales on the other, with the totals for each column having to match. Any remainder was either profit or loss. While modern accountants take this system for granted, it was a major innovation in the medieval period.

Early in the medieval period, money circulated in the form of coins. This led to practical problems for merchants and other businesspersons, the most obvious of which was transportation. Moving vast amounts of coin from place to place required high security and a means of transport. The cost of the process was frequently more than 10 percent of the entire value of the shipment. Medieval merchant banks offered a solution to this problem, because they dealt with paper rather than with piles of coin. These institutions were run by moneychangers who took care of all the physical moving and storage of coin, while allowing customers the ability to charge against their bank accounts; customers simply dealt with paper transactions. Often two parties to a business transaction had accounts at the same bank, and then the entire matter was a simple movement of numbers within the bank ledgers. Bills of exchange were an outcome of this new paper banking system. This instrument permitted the holder of the bill to receive money in one currency at one location and repay it at another time in a different location and currency. For international merchants, the ease of money movement expedited their trade. Another piece of economic paper, the letter of credit, removed the burden of carrying amounts of coins. Banks offered letters of credit to travelers such as pilgrims, students, and diplomats. A bank would sell the letter for a certain amount of money, which would be stated in the letter. The letter could be redeemed at the destination location (Hunt and Murray 1999, 63–66). Think of this as a medieval traveler's check.

The earliest mention of an institution resembling a modern bank is found among the Genoese moneychangers around the thirteenth century. They not only exchanged money but also loaned on credit and accepted deposits. Money deposited accumulated interest and had to be returned to the depositor within a specified number of days after it was requested. Banks in Venice had two types of deposits: regular and irregular. In a regular deposit, a customer entrusted a sealed container of valuables, sometimes money, to the bank for safekeeping. The bank was not permitted to use the goods but could only ensure their security of them in exchange for a fee. The irregular deposit was a monetary one. In this instance, the depositor did not expect the same money back, only an equal amount. The bank was free to use the depositor's money as it saw fit. However, once a withdrawal request was made, the same amount of money had to be returned (Mueller 1997, 10–11). The role of interest, both given (in the case of deposits) and taken (in the case of charges for a loan), had to skirt an official Catholic Church proscription regarding the sinfulness of the practice. Enterprising bankers, however, charged "fees" for loans and "rewarded" customers for their deposits. Thus, strictly speaking, no interest appeared on any account. By the fifteenth century, banks had expanded and had offices in all the major mercantile centers, such as Barcelona, Genoa, Milan, Pisa, Naples, and Venice. While the emphasis was still Italian, banking became common in other European localities, though these were often branch offices of banks based in Italy (Krirshner 1974, 202).

2

ARCHITECTURE AND CONSTRUCTION: HOMES AND CATHEDRALS

Perhaps the most impressive physical remains of medieval Europe are architectural. Castles, both intact and in ruin, are found scattered across the modern European landscape. They are stone testaments to a long-forgotten age. Even more impressive are the remains of massive churches built as receptacles in which to receive the glory of God. Romanesque and, later, Gothic church design illustrates the interconnectedness of technological know-how and piousness that are characteristic of medieval Europe. What is more, the stained-glass windows that covered the interior of late medieval churches with colored light were not only technological and artistic marvels but also a spiritual reinforcement for a society that sought higher connections with its God.

This chapter details medieval architecture and construction projects. At the heart of the medieval experience was the village. A description of the techniques used to build peasant houses opens the chapter, which then addresses the construction of the more substantial manor of the local lord for whom many peasants worked. While today chimneys are almost an invisible technology, in the medieval era they were innovative and sparked a change in house design. Castles were imposing reminders of the often-violent nature of the era, and their evolution and construction are examined next. Romanesque and Gothic churches were some of the most massive building projects in all medieval Europe, and each design type, with its own architectural devices, is considered in turn. Stained glass and the theological significance of light conclude the chapter.

VILLAGES AND MANOR HOUSES

A medieval village is best defined as a self-contained group of homes that together ensured the survival of the inhabitants through shared labor and, frequently, collective ownership of farm implements. The houses sat at the center of the village, with farmland and pastures radiating out from the central core. The manorial system of villages was one in which the villager-farmers worked on land belonging to the lord, who lived in a manor house near the farm. In this situation, the farmers were dependents of the lord, who may have owned several manors on each of his various land holdings. Variations of village layout did occur. Early medieval villages, in Germany for instance, took several forms: straight lines of homes along a roadside, or rounded villages with a common green space often set aside for a church. In areas near the Elbe River, German towns took a circular configuration as a defensive measure.

The type of house constructed by peasants depended greatly on their location within Europe. Peasants living in the warm climate of the Mediterranean did not have to worry about the drastic range in temperature from one season to the next and therefore built relatively simply houses designed to keep out rain rather than to provide shelter from the cold of winter. In contrast, northern European homes had to guard against a greater variety of weather and, as a result, were considerably more robust. In places like Germany, for example, peasant farmhouses were constructed out of wooden planks because, in areas like this, which had a great number of forests, wood was tremendously cheap. As the family grew in size (i.e., had more children), the house had to be altered to accommodate the new members. Archaeological evidence suggests that these houses underwent several additions before being replaced by larger structures. While early village houses were built directly on the ground and thus had dirt floors, the remains of some later villages reveal that houses sat upon stone slabs. The use of a slab allowed other innovations of house construction, such as the replacement of post construction, in which the supporting beams of the home were set into the ground, with frame constructions that sat on top of the new stone foundation. The advantage of the frame-built houses was their longevity. Where traditional post-style construction rarely lasted the lifetime of the owner, frame houses would survive for several generations. In various parts of medieval Europe, it was common for families and their animals to share the same living quarters. This arrangement, however, seems to have occurred with greater frequency in Scandinavian countries than it did in populations located more in the center of continental Europe (Rösener 1992).

Some peasant houses in England and other countries lagged behind their counterparts on the Continent in terms of technological sophistication. These houses were mostly dirt and unprocessed branches, with thatched roofs put together in what is known as wattle-and-daub style. This was a method of building walls from very simple materials. Walls were constructed by first placing wooden stakes—usually thin tree branches—into the ground. Next, green branches (they had to be green because suppleness was required) were woven around the stakes like a basket. The resulting lattice of branches is called a wattle. The walls were then covered with daub on either side. Daub is a mixture of mud and straw, with animal dung as a binding agent. Once the daub dried, it was frequently whitewashed (with a paint made from slaked lime, or heated limestone) to make the home more watertight. Windows were small and had no glass, only wooden shutters. The roofs of these houses were thatched with straw, which, when woven together, was quite waterproof. Floors were most often only the uncovered ground. Wattle-and-daub houses had at most two rooms, with one being much more common. When a peasant was wealthy enough to construct a home with multiple rooms, the overall size of the house was still very small, measuring approximately 15 by 15 feet for the entire family. Unlike some of the sturdier German peasant homes, these wattle-and-daub house were not permanent (Singman 1999, 82–84).

The manor house in medieval Europe was very important in the everyday life of most people, elite and nonelite. As the historian Mark Bailey puts it, "the manor touched and influenced the lives of the common people in the Middle Ages, probably more than any other secular institution" (Bailey 2002, 1–2). This was the case because the manor was the focal point for local administration. It was the physical connection between the lord and the peasants who farmed his fields and stood in sharp relief against the simple peasant homes. The early manor house consisted of a rectangular single-room building made first of timber and later of stone. This design became known as the manor hall and was modeled on the basilica shape of the parish church. The hall was the focal point of activity in the medieval manor house. Other necessities in the rectangular structure, such as kitchens and pantries, were at one end of the house, while the lord's personal room was at the opposite end. Floors in these first manor homes were generally dirt and straw. Only the dining area of the noble family was paved with stone. Artificial floor coverings were a later addition (Cook 1974, 10–12; Woolgar 1999, 47). Later manors had a second floor, accessed by exterior stairs, where the lord and his family slept. While primitive by modern standards, the manor house represented the pinnacle of house ownership in the medieval age. This is seen in the use of glass in large widows, a luxury not present in peasant homes.

While constructing the walls was a relatively simple matter of joining stone with mortar, interior ceilings were more complicated. Roof design in manors evolved from the "king-post" type to the more elaborate "hammer-beam" construction. The result was a dramatic increase in openness within the house. The king-post style of roof is characterized by a triangular support with a post (king-post) running straight from the apex of the triangle to the center of the horizontal support (think modern rafters). The hammer-beam style contains an open-air element and is very different. In this roof, where the structural aspects of the construction are completely open to constant visual inspection, form and function necessarily collide. The most dramatic example of this type of design is found in Gothic cathedrals. A "hammer-beam" was first used in England in the fourteenth century. The components originated in France and the Low Countries in the late twelfth and early thirteenth centuries. The historian of medieval architecture Lynn T. Courtenay describes "hammer-beam" roofs as those "in which the principle frame contains a horizontal beam (the hammer beam) perpendicular to the wall and projecting at wall-plate level beyond the vertical plane of the masonry" (Courtenay 1985, 90). The supports for the roof rest on the hammer-beam, thus creating a very high ceiling because there is no horizontal support spanning the entire distance between walls.

Prior to the end of the thirteenth century, the nobility in England tended to be rather nomadic, moving from one residence to another during the course of the year. After this time, manor houses were built to accommodate stays of longer duration, and the emphasis in architecture changed from defense to display. The change to single residences meant that the resources needed to maintain several manors could be diverted into the upkeep of one opulent manor, which, by this point, was a stone-and-masonry structure. One-upmanship became common among lords in the countryside, with everyone trying to outdo his neighbors. With less wandering from manor to manor, lords no longer required portable furniture and decorations. The result was lavish tapestries and beautiful heavy carved chairs and bed frames, to mention only a few of the manor's holdings (Woolgar 1999, 70).

CHIMNEYS

The chimney developed as a response to the drop in temperature beginning in the early twelfth century, an era known to historians as the Little Ice Age. During this time there were new icebergs in the sea, and formerly productive land froze. The temperature dropped only a few degrees during this era, but it was enough to drastically change the length of the growing seasons. Greenland, for example, lived up to its name in the tenth century. However, by the thirteenth century, the once lush fields took on their more modern appearance of fields of snow and ice. From 1350 to 1550, the climate warmed again. Prior to the invention of the chimney, a single fire in the great hall of the manor house, or in the center of a peasant's home, provided heat. Smoke from the fire escaped through a hole in the roof, although not before filling the entire room. The earliest English chimney was built at Conisborough Keep, in Yorkshire, around 1185. With the chimney came the flue, which directed sparks away from the interior of the home. It was now safe to have fires in other locations of a home and not only in the center of the main room.

Chimneys altered the design of buildings. Manor houses were being constructed with less wood and more stone so that a chimney could be built into an exterior wall with no fear of potential fires. The chimney itself demanded a wall built of brick as a fire retardant. Strong chimneys could also act as the backbone for second-floor constructions. And, since the chimney was already in place, a second-story fireplace meant that an additional floor could be heated. Lords of the manor moved upstairs because the second floor was warmer. The days of single great rooms and everyone living together were over. With heat now available in multiple areas of the house, rooms became private because members of the household were no longer required to huddle together for warmth around only a single fire. Fireplaces were introduced in the homes of the ultrarich in the thirteenth century and into those of the merely well-to-do by the fourteenth. Fireplaces also changed how food was cooked. By the fifteenth century, turbines in chimneys used a system of gears to power rotisseries that automatically turned the meat so that it did not burn (Burke 1978, 157–159).

Some historians have suggested that fireplaces and chimneys were aristocratic technologies that did not impact peasant families. It is argued that chimneys were absent from the houses of the poor, which continued to fill with smoke. The historian Werner Rösener has offered a different picture and notes the dramatic impact that the chimney and its companion, the heating oven, had on the development of peasant homes in medieval Germany. This new heated and smokeless room (called a *Stube*), dating from the twelfth century, became a focal point of peasant life in the coldest months. Because the stove in a *Stube* was fed from the back, which itself was outside the room, the *Stube* remained absolutely smoke free. Such an achievement was possible only with the use of a heating stove and chimney. Even in more aristocratic manors that relied simply on a fireplace with chimney for heat, the room still became smoky (Rösener 1992, 79–81).

CASTLE CONSTRUCTION

Early medieval castles were made of logs and earth and are known as "motte-and-bailey castles." The basic plan called for a wooden tower built on a hill (the motte), which was surrounded by a ditch that was often filled with water. This central tower would remain a common feature in medieval castles and is known as "the keep." Beside the motte was the bailey, a fenced area containing several buildings that frequently acted as a kind of sanctuary for local people. Despite the fact that these early castles—William the Conqueror built more than 70 of them—were signs of prestige, they tended to be only as strong as the nearest fire, the weapon of choice for attackers. The technology involved in constructing this kind of castle was not very sophisticated, usually consisting of little more than an axe to chop trees and shape the logs into building material. Labor for this work came from the peasants in the surrounding area.

By around the eleventh century, the economic turmoil of the previous decades had subsided, and local lords were keen to centralize the administration of their lands and to create tangible demonstrations of force against those who might oppose them. As castles became regular residences, they also took on a major role in the social life of the countryside. Defensive capabilities differentiate castles from manor houses, even though they performed many of the same tasks. In the lord's castle, one could expect to be entertained and to take care of any administrative business, such as paying taxes. Since wooden castles did not convey a sense of impenetrability, a new material was required. Castles were made of stone masonry and were intended to be permanent structures on the landscape of Europe. These structures date from the late tenth and early eleventh centuries. Two obvious advantages of this construction method leap to mind: stone castles were impervious to fire and were not prone to rot (Nicholson 2004, 77–78).

Remains of a stone castle keep.

However, construction of stone castles ushered in a host of problems not found in previous wooden designs, related chiefly to the difficulty of moving and preparing the vast amounts of materials needed. Consider the stones used to build a castle in the French region of Anjou: the medieval military historian Kelley DeVries notes that the walls sat upon a foundation of crushed rock about 2.3 feet below the ground surface. The castle walls themselves were assembled out of as many as 144,000 ashlar blocks (this means that the stones were made square prior to installation), with dimensions from four to five and a half feet thick on each side. The spaces between the stone blocks were filled with smaller stones and limestone mortar. Castle walls were thickest at the bottom and gradually tapered as they rose. The actual thickness varied from one castle to another, but rarely did they exceed the range of 3.2 to 6.5 yards. Wall stability, as DeVries explains, was achieved by its structure, by the substructure, and by "the fact that there were few openings in the walls" (DeVries 1992, 213–214).

Skilled masons were required to cut and place stones, which drove up the costs. The master mason served as both the architect and the general contractor on medieval job sites such as castle construction. These men (they were always men) were highly sought after and could name their own price. The historian Jean Gimpel describes the extravagant lifestyle of the thirteenth-century architect John of Gloucester, who owned five houses and was handsomely paid for his work: he was exempt "from certain taxes," paid "double his salary when travelling," and given "casks of wine" and two fur coats for himself and one for his wife (Gimpel 1976, 115). Stone, unlike wood, does not grow on trees, and obtaining enough materials for an entire castle could be extremely expensive if a castle was constructed any great distance from the quarry. Often the cost of transporting stone was more than the actual value of the stone.

The existing stones of medieval castles tell historians much about the techniques used in the construction process. For example, scaffold holes are still present. This archaeological evidence reveals that temporary structures were erected to serve as working platforms and for material storage as the castle wall took shape and became higher. Holes in the stones of towers indicate the use of spiral scaffolding. Further evidence points to the use of onsite kilns to manufacture limestone for mortar that would bind the prepared stone into walls. The tools used by masons and bricklayers in the construction of castles are remarkably modern in appearance. Such implements include hammers, augers, chisels, and trowels, to name only a few. More advanced tools, likely used by the master mason himself, include a plumb-bob to ensure straight lines and level walls (Kenyon 1990, 164). While the glory of the design went to the masons, the experience of lesser people in castle construction was likely limited to carrying heavy stones on their backs, both on level ground and up the wooden scaffolding. (Many of the techniques used to build castles were also used to construct the grand Romanesque churches.) The time it took to complete a castle was directly proportional to the immediate danger. For example, castles built by Crusaders in the Holy Land went up faster than any other castles in Europe. On lands that lay between warring kings, castles were built with great haste. This was the case in part of modern Belgium, which is full of impressive castles.

By the late thirteenth century, new economic realities were signaling a reduction in the importance of castles, which had been designed to protect a population of agricultural workers. Urbanization centralized many trades and crafts people into newly emerging towns and cities; castles in the countryside could not serve their defensive needs. Walled towns were the answer to the newly burgeoning medieval economy. Technological advances also demanded a change in castle design. Whereas the traditional wall, thicker at the base and gradually thinner at the top, was perfect for protection against cavalry assaults, the arrival of gunpowder weapons required a redesign. By the end of the fourteenth century, the defensive strategy still required high vertical walls, but now of increased thickness. The scholar Quentin Hughes argues that "Circular or rectangular towers tended to be replaced by pentagonal or oblique-sided ones to deflect shot" (Hughes 1974, 67). One technology drove the advancement of another.

ROMANESQUE CHURCHES

Some of the most impressive architecture in medieval Europe was inspired not by desires for protection and warfare, as in the case of castles, but rather by a pious duty to God and the goal of worship. This is surely the case with Romanesque churches. The term "Romanesque" is an invention of modern scholars and denotes the hybrid appearance of these churches: not quite Roman and not quite Byzantine. Romanesque churches were constructed mostly in the eleventh and twelfth centuries. Typical of the design were the square exterior towers and thick walls with round arches. These churches usually followed the traditional "basilica" shape of churches common in the late Roman Empire. Most simply, a basilica was a rectangular building with a flat wooden roof. These new churches abandoned wooden roofs, and this gave the churches their characteristic appearance. In Romanesque churches, the "nave," or central passage, had over it a tall and substantial round vault (called a barrel vault) ceiling made of stone and masonry elements. In later designs, a cross vault was used when two basilicas met at perpendicular angles to create a floor plan in the shape of the cross. A barrel-vaulted ceiling is fashioned from a series of rounded arches so that they form a complete covering. The stones in the roof and arches over doors and windows were first set in place on wooden supports that secured them until the mortar set. Once the mortar was dry, the supports were removed. The specific shapes of the stones for the arches was a trade secret among the masons, with the keystone (the stone at the pinnacle of the arch) being the most important because it directed the weight down the stones on either side of the arch rather than directly beneath it (Coldstream 1991, 40, 44–45). Because of the heavy ceiling, Romanesque design employed massively thick walls and great empty interiors, with very large pillars supporting the roof. As a result, there was very little space for windows, and the churches were somewhat dark. Nonetheless, the look of these massive stone churches became very popular and had international appeal. Examples of Romanesque churches are found in Germany, France, Spain, Italy, and England.

Romanesque churches represent a wonderful example of medieval construction projects. These churches also reveal a change in architecture style and ambition. The designs of most pre-Romanesque architecture copied those of existing buildings. While masons could cut stone with various decorative elements, thus making some structures superficially distinct from others, there was no major departure from earlier forms. Around 1050, architecture became more complex than in earlier periods. This was a result of design changes, rather than innovations in methods of material construction, chiefly stone masonry. The historians Charles M. Radding and William W. Clark suggest that the use of "more planned and

Example of round arch.

consciously innovative style of construction" in the eleventh century produced buildings in the style known as Romanesque. They claim, further, that the motivation of novelty lay in the "the ambition and originality of the designs that these builders conceived and executed. Simply stated, the shift is away from thinking of design in terms of flat, undifferentiated planes of walls and ceilings toward discovering means of delineating the spatial units and volumes contained within buildings" (Radding and Clark 1992, 12). These Romanesque structures were much larger than buildings of the ninth and tenth centuries and are indicative of a prosperous society that could afford to devote many resources to such large projects.

Design was nothing without the means to implement it, and this task fell to the masons who carved stones for and often supervised the construction of medieval buildings such as castles and churches. Most European masons were schooled on the job. After being trained, a mason became a journeyman and practiced his craft at several sites prior to undertaking a project of his own. In this respect, being a mason was the same as participating in any other medieval trade. Very often, a particularly sought-after mason would supervise several projects at once. In addition to the usual activities of stone cutting and planning, it was not uncommon for

Example of medieval mason construction in a church.

enterprising masons to branch out into other aspects of construction, such as quarry ownership and stone transport. Although this multitasking was relatively unusual, it does reveal the central role that the masons played in the medieval building process (Coldstream 1991, 15, 17). However, it was rare for the original job-site mason to be the overseer of the final stages of construction. Given the very long time spans needed to complete cathedrals or similar structures, it was not uncommon for at least one entire generation (very often more) of workers to have died and been replaced as the project proceeded.

The center of activity of the medieval workplace was the masons' lodge, where the stone was carved and architectural plans made. While most stone used in the construction was square and did not require cutters of great skill for shaping, other stones were more decorative, and still others needed to be cut into precise forms. Masons transferred the design for the stone from templates that they had created prior to cutting any stone. These templates could be made from wood, canvas, or thick paper. Several tools such as saws, drills, and chisels were used to cut and shape the stones. Aside from these, the most important tool at the job site was the sharpening stone. The choice of stone depended on the locale of the building. While many different types of stone could be transported, the cost was so prohibitive that masons usually relied on the stone immediately available. A medieval construction site was very modern in its appearance. Consider the description of the environment by the historian Nicola Coldstream: "Permeated with the dust of lime mortar and the residues of cutting stone and wood, it rang to the sound of hammers, chisels, axes, and saws, as well as the shouts of the workforce as they maneuvered materials and machines into position" (Coldstream 2002, 83).

Construction on major projects like Romanesque and Gothic churches began with laying the foundation, which was set deep into the ground so that it rested on bedrock if possible and wooden pilings if necessary. Once the foundation was secure, the outer structure was erected before workers moved on to the interior details. The task of hoisting rocks up the wall-in-progress fell to less skilled workers who sat at the bottom of the medieval social hierarchy. Stones could be either carried up a series of scaffoldings or lifted into position using cranes. Cranes, while a technological achievement, also reflect the harsh social conditions of the medieval era. As in today's modern construction projects, a medieval crane was built on site and often on the building itself. Its function was the same as that of today's cranes: to lift heavy objects from one location to another and from one level to another by means of a cable that extended and retracted. Where modern and medieval cranes deviate is in the source of power used to make them work. Medieval cranes employed people to wind and unwind the cable. The cable was spooled around a shaft that extended into what is best described as a human-size hamster wheel. The shaft rotated as a person (sometime more than one person on very large cranes) walked either forward or backward inside the wheel. Because of the precarious position of cranes on unfinished structures and because the passing rungs of the wheel induced dizziness, it was common for workers to become ill. Blind men, who were otherwise seen as a drain on society, were the solution because they did not see where they were working and were not bothered by visual distractions.

GOTHIC CATHEDRALS

Like the name "Romanesque," the designation "Gothic" in reference to cathedrals built in the late twelfth and thirteenth centuries is assigned by modern historians and was not used in the medieval period. Whereas Romanesque design could be described as fortress-like, with its imposing stone structures, Gothic architecture is characterized by thin walls,

Depiction of medieval crane and construction. The Pierpont Morgan Library, New York. MS M. 638 f. 3r.

often filled with stained-glass windows that offer a feeling of airiness not found in Romanesque churches. To achieve this drastically different vision, Gothic design employed distinct architectural techniques.

The great age of Gothic churches was the thirteenth century. However, historians agree that the first example of a Gothic church was the abbey church of Saint-Denis, near Paris, which was begun around 1137 and completed sometime between 1140 and 1150. Abbot Suger (1125–1151) of Saint-Denis envisioned the design, which attempted to unite architecture and religious experience. Suger decided to rebuild the church of his abbey around 1124 in part to accommodate the increasing number of pilgrims who wished to view the relics of Christ's Passion that were held there. He was particularly keen to redecorate the entrance of the church, what is now the famed west façade of St. Denis. Sugar claimed his only inspiration for St. Denis was Solomon's Temple because both that project and Suger's were directed by God (von Simson 1988, 91–92, 95). Although it started in France, the Gothic style soon spread to England, Spain, Germany, and, indeed, to almost the whole of Europe.

Through a combination of ribbed vaults in the ceiling and pointed, rather than rounded, arches, Gothic churches could be constructed higher than earlier churches. The ribbed vault differed markedly from the barrel

Interior of a Gothic cathedral.

Ruins of a Gothic cathedral.

vault with its sequence of rounded arches constituting the roof structure. Ribbed vaults were built from masonry that formed a skeleton support of stone for the roof. The roof, made up of stone blocks, rested on top of the intersecting ribs, which were often visible on the completed structure. Thus, the ribs served both an aesthetic and an architectural purpose. However, until the mortar holding the roof stones set, ribs were the main support in the ceiling. After the mortar dried, the ribs were purely a decorative element. This design allowed the roofs of Gothic church to be much taller than those of earlier buildings, because the weight of the ceiling was considerably lighter and therefore less complicated to support. Pointed arches also permitted Gothic design to tower over its Romanesque predecessor. An arch is supported by the slight wedge shape of its individual stones, especially the top (or key) stone. As churches grew ever taller, the round arch could no longer accommodate such increases because, when rounded arches are used, the weight of the roof (the force of gravity pulling downward from the center of the arch) is too great and the arches simply begin to spread apart because their shape directs the force both vertically and horizontally. Pointed arches, because of the triangular shape of their keystones, direct the force of gravity straight down only. What is more, the downward force of gravity pulling on the keystone is balanced by the upward force of the stones on either side of it. The equal forces ensure that the ceiling remains standing.

The most important innovation in Gothic churches also made the greatest visual impact. Flying buttresses allowed the weight of the ceiling to be distributed outward and downward onto the external buttresses rather than onto the walls, as in Romanesque designs. This allowed for lighter roof supports, such as the ribbed vault. Basically an arched stone pillar extending outward from the church, the flying buttress allowed the walls in Gothic churches to be much thinner than those found in their Romanesque counterparts. With the ceiling support coming from outside

Rib vault construction in a ceiling.

the church, large portions of the Gothic walls could be used to display stained-glass windows. As Otto von Simson has put it, in an age where learning was described as illumination, the walls of Gothic churches, more glass than wall, continued this metaphor. Historians praise the flying buttress as a major technological creation. For example, the famed scholar Jean Gimpel writes, "Flying buttresses were one of the great 'inventions' of the medieval architect-engineer. They were an ingenious and revolutionary building technique, conceived to help solve the technical problem created by the desire to give maximum light to the churches while raising the vaults higher and higher" (Gimpel 1976, 121).

As walls in Gothic churches grew ever higher, new problems, such as wind sheer, became evident. Again, the flying buttress solved this issue. This is apparent in the Cathedral of Notre-Dame de Paris, started in 1160, which was eight and three-quarter yards taller than any previous church. It was here that the flying buttress was first introduced as a design element during the construction phase around 1170. The impressive flying buttresses on the church were more than weight bearing; they helped keep the tall, thin structure standing even in the strongest windstorm. In the words of Robert Mark and Huang Yun-Sheng, who conducted scientific modeling of medieval churches, "the role of the upper flying buttresses in the classical High Gothic … acts mainly to resist high wind loadings on the church superstructure" (Mark and Yun-Sheng 1985, 137). In spite of their radically different appearance, the construction methods and material used in Gothic and Romanesque churches are very similar: masonry and stone held together with limestone mortar.

GLASS, STAINED GLASS, THE SCIENCE AND THEOLOGY OF LIGHT

The recipe for producing glass was already very old at the start of the medieval era. Glass is made from a mixture of silica, sodium, and calcium carbonate. One of the earliest recipes for its manufacture is found on a Mesopotamian clay tablet dating from between the fourteenth and the twelfth century B.C.E. If nothing was added to the ingredients, the resulting glass tended to be bluish-green because of the iron content in most sand. Colored glass was made by adding certain metals: copper made pale blue, dark green, or various reds, cobalt resulted in glass that was deep blue, and manganese made yellowish glass. Almost clear glass was produced by the selection of silver-colored sand, meaning that it was free of iron.

Around the middle of the sixteenth century B.C.E., the first glass vessels were made. Glass was initially used to copy objects, such as goblets, that already existed in different materials. The Greeks were making yellow and green glass in the fourth and fifth centuries B.C.E. It was during the Hellenistic period (from the death of Alexander the Great until the establishment of the Roman Empire, ca. 323 to 31 B.C.E.) that blowing emerged as the dominant glassmaking technique and superseded previous methods, which tended to focus on using forms and molding glass around them. By 50 C.E., glassblowing had eliminated other techniques entirely. Roman artisans had glass-making and glass-blowing factories in Rome and in Campania, where they made bowls and other vessels. After the Roman Empire fell, the Franks (a Germanic tribe that occupied much of modern Germany and France) continued the glass-making tradition. Other Germanic tribes, the Angles and Saxons, went to England after the Romans left in 410. In the ninth and tenth centuries, English glass-making seems to have been restricted to the north and the southwest of the country. Around this same time, monasteries in northwestern Europe were using glass windows, although the practice undoubtedly began earlier. In England, around 675, Benedict Biscop, founder of a monastery in Northumbria, solicited glassmakers from France to make windows for the building. Windows have been discovered in excavations of churches mostly in northern Britain, as well as in Winchester, toward the south (Tait 2004).

The process of producing a flat piece of glass suitable for a window required a tremendously skilled craftsman. There were two methods of making glass for windows in medieval Europe. The first is known as the "muff" method. In this technique, the glassmaker collects a sphere of red-hot glass on the end of a hollow iron rod known as a pontil. The hot ball is then rotated to mold it. Next, the glassmaker blows the glass into an elongated oval shape (called a muff) and cuts it at both ends. Finally, the muff is cut lengthways and flattened with a wooden tool that resembles a cooking spatula. The second method is referred to as the "sheet or crown" method; the initial steps are the same as in the muff method. As soon as the molten ball of glass has been rotated, a second pontil is attached to the other side of the ball and the first pontil is removed. The ball is spun to expand its size. During the spinning, the hot glass gets thinner and flatter. When the desired size has been reached, the process stops, with the result being a flat, circular piece of glass that, when cut to size, is suitable for windows (Brisac 1986, 180–181).

The most famous windows of the medieval period were the stained-glass windows, containing pictures of saints and biblical stories, found in churches and cathedrals. Early examples of stained glass were likely decorative only and used smaller pieces of flat glass. The development of stained glass is closely linked to the changing style of church architecture. Windows could be only as large as building designs permitted. As a result, the original windows, likely dating from the first century C.E., were quite small. This trend continued well into the twelfth century, through the Romanesque period, because the entire function of walls was to hold up the roof and that prohibited the replacement of large wall segments with glass. However, the thirteenth century, as Catherine Brisac has noted, is "quite rightly regarded as the heyday of stained glass." New architectural designs allowed for thinner walls and larger windows. This was the epitome of Gothic construction. Glass could now fill spaces that were formerly needed for the structural integrity of the church. One of the most noted examples is Chartres Cathedral, which contains around 160 separate stained-glass windows (Brisac 1986, 33).

Producing stained glass took the talents of skilled artisans and craftsmen. Properly speaking, stained glass is more accurately called painted glass, because the specially trained painters (glass-painters) used glazes to create their masterpieces, which were then fired to set the image onto the glass surface. Nonetheless, "stained glass" is the more common term and is used interchangeably with "painted glass" in this section. Despite numerous colorful reminders of their work, most medieval glass-painters toiled in historical obscurity. We know next to nothing about the craftsmen who created some of the most lasting and most radiant monuments of their era. Medieval glass-painters were part of cathedral construction teams, along with masons, carpenters, and countless laborers. They were professional artisans who, in addition to creating their own masterpieces, were often obliged to repair existing windows. Becoming a master glass-painter, as the historians Sarah Brown and David O'Connor have revealed, required a major time commitment. As they explain, "Training, which seems usually to have begun at the age of ten, was given only in the milieu of the workshop and lasted for a number of years—four, seven or even ten." Then an apprentice had to prove himself with a "master-piece" example to demonstrate his readiness to become a master in his trade (Brown and O'Connor 1991, 23–24).

Much of our information regarding the craft of making stained glass comes from the accounts preserved by Theophilus, a Benedictine monk who wrote in the twelfth century. While Theophilus included glassmaking in his description of stained glass, it was extremely rare for glass-painters to make their own glass; rather, they purchased prefabricated pieces from merchants. The design was first set out on a planning table, and the required cuts and painting were calculated. Both colored and clear glass were then cut with hot irons. This technique was replaced in the fourteenth century by the use of a diamond tip cutter. Having created the necessary pieces, the painting took place. The task required special paint composed of copper or iron oxide, "frit" (finely crushed glass that allowed for more rapid melting), and a binding agent (often wine and sometimes urine). Usually, three coats of paint were needed. Each area in Europe developed its own trademark painted glass. For example, craftsmen in Strasbourg around 1260 were known for panels depicting a single figure enclosed within a building-like structure. Once painted, the pieces were fired in the kiln at about 1100 degrees Fahrenheit. After the pieces had cooled, they were returned to the planning table to be joined with lead that was shaped around the piece of glass and soldered.

While they were pretty to look at, these colorful scenes of religious life were not simply decorative. They were created in praise of God. Light was associated with the glory of God. Bishop Durand de Mende stated around 1300 that "Stained-glass windows are divine writings that spread the clarity of the true sun, who is God, through the church, that is to say, through the heart of the faithful bringing them true enlightenment." Colored glass turned the interior of churches into a venue bathed in colored light—an almost mystical testament to the glory and mystery of God. Thus, elite and nonelite churchgoers were literally covered in the colored rays of God. On this association, the medieval observer Pierre de Roissy wrote, "The stained-glass windows that are in churches and through which ... the clarity of the sun is transmitted, signify the Holy scriptures, which banish evil from us and enlighten our being" (Brisac 1986, 7, 13). Not all historians accept the prominence given to light when discussing the importance of stained-glass windows in everyday religious experience. Wolfgang Kemp argues that the windows often told a narrative tale of moral or theological significance through pictorial representation of the story elements. With the majority of Europeans not able to read, these pictures provided some of their lessons in theology. Kemp asserts that this function of the windows takes precedence over the notion that bathing the cathedral in colored light was the most important task of the stained glass (Kemp 1997).

For Franciscan friars, the contemplation of light had both theological and scientific significance. As the historians Roger French and Andrew Cunningham have explained the medieval motivation for this: "*visible* light is the visible, earthly counterpart of *spiritual* Light. Spiritual Light is what God is, and which acts on the level of the intelligible. Visible light is what He uses to carry out His purposes in the sensible world. Study of visible light therefore tells one most directly about God and His actions." When Franciscan scholars such as Robert Grosseteste (c. 1168–1253) studied light, a topic about which Grosseteste was obsessed, they did so with the double goal of praising God and understanding the creation. Grosseteste believed that light was responsible for all change in creation. What is more, he believed that light was God's medium of action in the world, indeed God Himself; he argued that the creation began as a single point of light that expanded to fill the universe. From studying light, Franciscans progressed to contemplating the rainbow, studying how the eye works, and tracing the paths of rays of light (French and Cunningham 1996, 224, 231–232, 235). This emphasizes that medieval studies of nature were done not for their own sakes but to understand God and what God had created.

Much of the stained glass produced in medieval Europe was lost during the sixteenth century through the enthusiastic destruction led

Stained-glass window.

by Reformation-era iconoclasts who viewed stained glass as a form of idolatry. This was especially true in England, when the reforms of Henry VIII, beginning in 1543, ensured that much of the richness of English stained glass was reduced to broken shards.

3

TRANSPORTATION AND TRAVEL

Like their counterparts in the modern world, medieval people did not remain all their lives in one place. Whether for daily chores or religious experience or as part of their employment duties, people moved about their communities and around all of Europe. Products, too, moved where market demand was greatest. Merchants eager to take advantage of European fascination with silk and exotic spices had to travel to obtain this merchandise and then return to their home. This chapter outlines some of the issues and technologies involved in moving people and goods in the medieval period. Everyone in this time walked, and thus journeys taken by foot open the chapter. For those who could afford it, travel by horse permitted them to traverse great distances. Horse ownership was a mark of high social standing and was out of the reach of practically everyone in medieval Europe. The technology of riding—saddle and stirrups—are considered in relation to this. Animal power was also used to pull a variety of carts and wagons, which is the next topic. Whether walking, riding a horse, or traveling with a cart, people of the period experienced many kinds of roads. In many cases, the construction of these roads predated medieval society and could be traced back to the Roman Empire. The robust roads needed to control an empire survived into the medieval age and formed much, though by no means all, of the era's roads. Travel on roads was often precarious, and we discuss some of the everyday dangers. Roads can go only so far across the landscape, and where water interrupted the travel surface, bridges were built. Medieval construction produced many bridges of different types that are described here. Very rarely did cargo travel exclusively on land, and when particularly heavy items such as building stone had to be moved, waterways were employed. Both natural and artificial canals were used extensively in medieval Europe. Water transport required specialized technology such as locks, the working of which is discussed. Last, transport ships of both the ocean-going variety and river barges are described, as are the international trade routes that covered much of Europe.

ON FOOT (PILGRIMS)

The most basic transportation technology of the medieval era was the foot. In an age when individual horse ownership was very expensive and usually reserved as a mark of aristocratic standing, the majority of people walked everywhere they went. This included trips of both short and very long duration. Modern estimates suggest that the maximum distance a person could have covered in a single day was around 20 miles. The study of walking and trip taking is made somewhat confusing by the specifically medieval use of recognizable modern terms. The historian Jean Verdon explains the issue: in medieval Europe, the word "journey" was taken to refer to the road. It came to mean "movement of pilgrims" and then to refer specifically to militant pilgrims, that is, Crusaders. To take a journey was to visit a holy site or defend the Holy Land by participating in one of several crusades. Verdon continues: "in the fifteenth century, the term 'to journey' was understood to mean to make military expeditions. It was only in the late Middle Ages that 'journey' and 'travel' began to take on the meaning with which we are familiar" (Verdon 2003, 1). A pilgrimage demanded walking because it was seen as introducing a kind of penance into the journey. Often one had to walk in bare feet as part of the process. This occurred in 1095 when Pope Urban II urged that a crusade be organized to take the Holy Land back from the Turks; Peter the Hermit, who inspired a group of peasants to take up the task, first answered the call. The "Peasants' Crusade" was no match for the well-trained Turkish forces.

Those who did not go barefoot, either voluntarily as a form of religious devotion or out of necessity because of extreme poverty, wore simple shoes. These shoes were made from leather, including the flat sole. While more elaborate coverage of the foot was common on wealthy people, the rest of society wore a shoe that laced around the back of the heel (Singman 1999, 42–43). An ankle-boot style was very common. The manufacturing technique employed in medieval shoes is called "turned shoes." This means that the shoe was sewn inside out and then turned right side out (think of moccasins). While not very sophisticated and capable of producing a shoe with a soft sole only, the method remained in use a long time. Other methods included creating an upper part and attaching it to the sole with woolen thread or rawhide, depending on the European location in which the shoe was produced. Eventually leather thonging became the binding material of choice. The advantage with this style of shoe was that a worn sole could be replaced and the upper portion of the shoe used again. Buckles and shoelaces as methods of fastening appear on shoes dating from the seventh century. Shoe style changed as fashion demanded. In the middle of the twelfth century, a brief fad for extremely pointed toes came and went. By the fifteenth century, a technique called "turn-welt" allowed thicker soles to be attached to shoes, increasing both the longevity of the shoe and the amount of protection it offered the foot.

While most people never traveled much more than a few miles from their village, soldiers and merchants did. But, even then, fewer than 30 percent of them went very far from their homes. Other members of society, too, traveled extensively. From friars preaching the word of God to bishops and other Church officials carrying out visitations, members of clergy traveled a lot. Diplomats and royal agents traveled for business and to issue proclamations.

HORSES

While the majority of medieval people walked, the wealthiest rode horses. A horse cost as much as two simple peasant wattle-and-daub homes and large warhorses even more than that. A person riding a horse could cover about 30 to 35 miles per day, which was nearly double the distance that could be traversed by walking. At the highest level of society, the horse was an obvious visual display of wealth and signaled its owner's military activity as a mounted knight. An armored knight and horse were an imposing battlefield tool, and the combination has been described as a "moving castle." Although horse ownership was a crucial ticket of entry into the aristocracy, how one handled one's horse was just as important as the possession of one. The long hours needed to become a proficient horseman meant that one had to be free from other duties of life, such as farming. What is more, medieval lords tended to travel a great deal to visit their various land holdings, and thus horse riding was part of their everyday life (Ayton 1994, 25, 32–33).

For the rest of society, the horse was an occasional (at most) form of transportation and usually the only alternative to walking. Thus, less wealthy travelers might still use horses but not as mounts. Rather, horses were used to carry cargo either in saddlebag-style baskets or by pulling a cart, while the owner walked beside the horse. More often than not, though, horses were too expensive to be traveling tools, and instead donkeys served as pack animals. Horses were needed in the field to pull plows (Singman 1999, 214–215). For those who did not walk, the mule or donkey was a common method of transportation.

Until the late Roman Empire and the early medieval period, horseback riding was done without modern equipment such as a saddle.

Mounted riders sat upon a variety of blankets. The enormous athletic skill needed to ride bareback both into battle and for transportation was rendered somewhat unnecessary with the invention of the rigid saddle and the stirrup (Piggott 1992, 74, 89). Saddles evolved from a simple blanket to a pillow (stuffed ironically with horsehair). Riders noticed that placing all their weight directly on the horse's spine caused sores and resulted in a lame mount. The technological innovation that solved this problem combined knowledge of horse anatomy and leather working and resulted in a saddle that placed pads on either side of the spine and then joined them to form a leather seat. Rigid backs on saddles were a later invention that resulted from the experience of knights in battle who tended to fall off their horse after the impact of jousting (Brereton 1976, 49).

In the medieval period, the stirrup first appeared in the ninth century and is represented in the Bayeux Tapestry's depiction of the Norman Conquest of England in 1066. As with so many other pieces of medieval technology, the roots of the stirrup may be traced to China around the fifth century. From China, the stirrup then appeared in Korea, Japan, and Muslim countries by about 700. Stirrups were usually made of iron and were shaped in half-circles or either semi-triangles. They were joined to leather straps extending down from the saddle by means of stirrup mounts made from an alloy of copper (for Anglo-Saxon-era England) or another metal. The decorations on the mounts suggest a high level of metalworking sophistication even in the late eleventh century (Williams 1997). The importance of stirrups is that they give the rider an increased ability to control and direct the horse. In addition, stirrups make it much easier to mount a horse. When combined with the bridle (the oldest piece of horse-riding equipment), the rider has more points of contact and is more securely attached to the horse. This was a major concern for knights who fought from horseback. It is surprising, then, that the stirrup was invented more than a millennium after the first bridle and early saddles (Brereton 1976, 51).

In 1962, the famed historian of technology Lynn White, Jr., argued that the introduction of the stirrup into Europe allowed the full capacity of the horse, as an instrument of war, to be realized. In his words: "The stirrup thus replaced human energy with animal power, and immensely increased the warrior's ability to damage his enemy." This military advantage was also effective domestically and became the backbone of the feudal system in medieval Europe by creating a class of warriors upon whom the rest of society depended for protection (White 1962, 1–38). White's identification of the stirrup as a piece of technology that brought major social change to medieval Europe has been challenged and refuted by more recent scholars, who attribute change in medieval society to a wider variety of technologies associated with the horse, such as the plow, collar, and systems of crop rotation. In this analysis, change is driven by agriculture rather than military conquest. Nonetheless, most historians remain wary of technological, deterministic explanations for changes in medieval society (Langdon 1986, 288–289).

CARTS AND WAGONS

When it came to transportation, carts (two-wheeled vehicles) and wagons (four-wheeled vehicles) were even more efficient than animals. The historian Jeffrey L. Singman writes that "a four-wheeled wagon with a pair of horses could haul as much as 1,300 pounds about 25 miles a day" (Singman 1999, 215). The earliest carts can be traced to the sleds used from about 7000 B.C.E. by the native tribes in eastern and central Europe.

The rise of wheeled transport around 5000–3000 B.C.E. took place where sufficient supplies of wood existed and where the technology existed to cut and shape the wood into wheels. Initially, wheels were solid rather than spoked. Wheels were shaped from flat pieces of wood. Wheels could be a solid piece of wood or formed from two or three pieces, depending on the size of the original tree and the desired size of the final wheel. Only basic tools such as an axe were required to produce wheels. Whether the wheel or the axle (in the form of logs placed under a heavy object to help roll them) came first is still debated. However, the combination of two wheels joined by an axle proved extremely efficient. Early designs had fixed axles with wheels, lubricated by animal fat and assorted oils, that rotated on them. Iron bands, replacing leather coverings, placed around the wheels improved the longevity of the wood (Lay 1992, 27–28). Locomotion came from animals, and carts pulled by oxen were common in Egypt, Greece, and Rome. This means of power was used into the medieval period. The major change, however, was that the prestige of oxen carts declined, and they were used only by those far down the social ladder, certainly beneath the rank of nobles. Medieval nobility and royalty turned to horses to pull their carts, while the rest of society relied upon oxen power for much of the early medieval period until they, too, took advantage of horsepower (Piggott 1992, 16–17, 28).

The use of wagons and carts in the medieval era, as the historian John Langdon has demonstrated, was widespread and cut across the economic spectrum. Carts pulled by horses were different from those destined to be pulled by oxen. Prior to the twelfth century, the majority of heavy hauling was done by teams of oxen; after this time, horses became the dominant service animals. The switch to horsepower, however, necessitated a reduction in the size and capacity of carts. Horse-pulled carts had roughly half the capacity of those pulled by oxen, which were much more robust animals. Where a particular cargo was especially heavy, such as timber or coal, ox power remained the preferred method of transport.

Both horse and oxen vehicles had, for the most part, two wheels. There did exist some four-wheeled wagons, but these were often reserved for road transport and then when the size of the cargo required it.

Oxen cart. The Pierpont Morgan Library, New York. MS H. 8 f. 4v.

The problem with having four wheels was that, even though pivoting front axles had been invented around 500 B.C.E, the technology did not spread very quickly. Early on in the medieval period, pivoting front axles did not exist, thus making turning a major exercise in animal control. While altering direction was difficult, it was not impossible; the fixed axle design lasted for 1,000 years in central Asia (Astill and Grant 1988, 91–94). Wagons with pivoting axles would not be common in England until the seventeenth century. Driving or riding in a medieval cart or wagon would have been a very bumpy experience because the vehicles had no suspension until the fourteenth century, when some carriage suspension was made from iron springs.

A less obvious example of a medieval cart used for transportation is the wheelbarrow. The wheelbarrow is of medieval origin and has no Roman predecessor. It is also unique in that it was the only vehicle of the time that was pushed rather than pulled. There are pictorial representations of wheelbarrows from the thirteenth century, although the technology certainly predates this (Leighton 1972, 88–89).

ROADS: CONSTRUCTION AND DAILY EXPERIENCE

Medieval Europe inherited the greatest system of roads the world had ever seen. Roman roadways were an engineering marvel that were most often used as military tools to move men and machines by the most efficient, if not always the most scenic, route. These roads were straight and had mileage markers and rest stops. Romans were prodigious road builders, and many examples of their craft still exist today. Economic necessity was the inspiration. As Donald Hill, a historian of engineering, reveals: "In general, the permanent roads were not built until a newly-conquered region had been fully pacified, and they were intended to facilitate the needs of administrators, tax-collectors, and merchants, although their use by the military was always important ." The most famous of Rome's roads was the Appian Way, started in 312 B.C.E. At the peak of empire, Rome could boast about 50,000 miles of roads. Construction of the roads was a project that involved many processes. Paving materials were taken from the locality in which the road was constructed, although some stone may have been transported over short distances. The road were paved with a series of layers. On the very bottom was placed a base consisting of sand covered by a layer of mortar made from limestone, for a total depth of about 8 to 12 inches. Next, crushed stones cemented with a limestone mortar (with a total depth of 12 to 19 inches) were placed over the base. Then, a third layer of concrete measuring 12 to 14 inches was added. Finally, the top surface was fashioned out of stones about 12 by 40 inches in size, which sat on yet another layer of mortar about 8 to 12 inches thick. The total depth of a quality Roman road ranged between 3.2 and 4.8 feet (Hill 1984, 81, 82–83).

Many Roman roads lasted long into the medieval period, while others, along passages no longer in use, were allowed to become overgrown. Caring for roads was often very low on the list of medieval landowners' concerns, and, as a result, the era of finely paved wide roads faded from the landscape of Europe for a generation or more. Some Roman roadways were plundered for their valuable paving stone, which became the foundation for new buildings. The development of new medieval industries not located along established Roman roads necessitated the paving of a new batch of roads so that their owners could access windmills and waterwheels and sources of salt. Moreover, the construction of great cathedrals created a demand for new roads that could sustain the transportation of the massive amounts of stone building material. In this new burst of roadwork, the elaborate substructure of Roman roads was abandoned. Medieval versions tended to be set on top of the natural soil (Leighton 1972, 53–55, 58–59).

Other concerns, too, made Roman roads obsolete. As one historian explains, the roads "were abandoned for the most part in the 10th and 11th centuries. A new network was put in place, created by private initiative. The purpose was no longer to cover distances as quickly as possible, but to connect all inhabited spots with one another" (Verdon 2003, 19). From the 1150s to the 1300s, new road systems were constructed to serve France's expanding economy, for example. Reports exist from around 1200 of the first medieval paving project. Also important to the growth of road systems was the political and administrative importance of Paris, to and from which many people had to travel. The increased use of horses also

Worn stones of a medieval road.

necessitated new roads. This meant new north-south roads, as opposed to
the Roman roads, which generally ran east to west (all roads led to Rome).
The historian Jean Verdon has identified five distinct types of roads that
might have been traveled upon in the medieval era: (1) the "path," which
was about four feet wide and permitted "travel from one major road to
another"; (2) the "cart road," which measured about 8 feet wide and on
which "two carts could not travel next to one another, but could pass one
another"; (3) the "way," which was around 16 feet wide and on which two
carts could travel side by side, cattle could be moved, and people could
travel from rural towns to castles; (4) the "road," which was about 31 feet
wide and upon which "animals have the right to graze and to stop, and
merchandise may go through; here, transport taxes are collected"; and
(5) "the royal road," which was the widest of all medieval roads at 63 feet
and whose enormous width ensured that "all the products of the earth
and the animals, on which men and women feed themselves to live, might
be led and transported" (Verdon 2003, 25).

Inside a medieval city, as in cities of today, streets defined the space.
However, medieval streets were narrow, unlit at night, and often muddy.
The smell would have been overpowering because, in an age before sanitation, refuse—human and animal—ran down a channel cut into the
center of the street. Paving was too expensive to be widely used, and dirt

streets were the norm. The constant passage of carts and animals turned the streets into soggy, rutted messes. Luxuries such as two-lane streets did not exist in medieval cities. Most streets were so narrow that they might be best described as walkways. Determining location by the geography of street names was not the simple exercise that modern travelers take for granted. Although medieval streets did have names, they had no signposts, and buildings had no numbers. Navigation was done by landmarks (Lepage 2002, 282).

The daily experience on medieval roads was often hazardous both financially and physically. Roads were the property of the lord over whose land they passed. European royalty was, from about the thirteenth century, keen on building roads for the revenue that tolls and taxes would produce. Thus, tolls on medieval roads added to the burden of transportation and travel. Aside from the monetary demands placed by some of those who owned the land over which the road passed, some tolls and laws were extremely harsh. In parts of the area covered by modern Germany, anything that fell from a cart became the property of the landowner. If a wheel came off a cart and the axle touched ground, the cart and all its contents became the property of the landowner. Thus, careful and lasting construction became an economical necessity in cart manufacturing. What made matters worse for those traveling with their wares or belongings was that some unscrupulous landowners were known to place hidden obstacles in roads to ensure timely mishaps (Leighton 1972, 94). While the tolls were, in theory, collected to pay for the maintenance of roads, many lords became quite attached to this source of revenue and viewed it as part of their entitlement. Dramatically increasing tolls became such as issue that, in 1179, a Church council stated that anyone who charged tolls without the permission of a king or a prince risked excommunication. Roads were often the sites of many dangers. It was not uncommon for travelers to be robbed, assaulted, generally harassed, and even killed. Bands of brigands (bandits) wandered well-traveled roads looking for easy marks.

Weather, too, affected roads. Winter turned roads into muddy bogs, and ice made horse travel very difficult. Mountain crossing was difficult, especially for those traveling with a large group. Snow and ice ensured that only the surefooted made quick progress. Crossing rivers was made easier with bridges and ferry services. Washed-out bridges were sometimes replaced by a row of boats. For those who traveled to the Holy Land from the northern regions of Europe, deserts were a new experience with their own set of challenges.

BRIDGES

Much early medieval river or stream crossing was done at shallow sections of the water known as a ford in a process known as fording. Where such areas did not exist, bridge construction was the solution.

Bridge building started in China and Asia much earlier than it did in Europe. In the medieval period, there were three basic types of bridges used: beam, pontoon, and arch.

The beam bridge style is the oldest and simplest to construct. In its most basic form, the traveler could place a log or a large stone in a small river or stream to allow the traveler to cross the stream while remaining dry. Where a large distance needed to be spanned, a beam would be set across natural islands in the river that would act as piers. In cases where no natural island existed, artificial piers were constructed so that a single beam did not have to join two shores on its own. The majority of bridges encountered by most people in their everyday lives would have been wooden beam bridges. Because of the structural limitations of wood, the longest span that could be covered with a single beam was around 20 feet. The center of the bridge was particular susceptible to bending. What is more, wood has a fairly short life span and must be repaired or replaced frequently. Examples from antiquity reveal that the length of stone beam bridges could be as great as 20 yards. These simple bridges were very common in everyday life throughout Europe until the end of the medieval period (Hill 1984, 64; Lay 1992, 254–255).

The pontoon bridge (bridge of boats) was an "important and widely used method of crossing rivers in the classical and medieval period, especially in the Muslim world." This kind of bridge may have originated when ships were moored side by side and boards were placed over them to making visiting other ships easier. These bridges can be constructed very quickly, especially in calm water. In rough water, a cable must be secured to both banks and the ships attached to it. Pontoon bridges were commonplace in Iraq, specifically in Baghdad, and used for crossing the Tigris. In other places, the pontoons (boats) served as supportive piers for temporary beam bridges.

The most celebrated medieval bridge is the stone arch bridge. Greek builders did not employ the arch to any great extent. The arch is "almost the hallmark of Roman civilisation" (Hill 1984, 72). The lasting remains of so many Roman arched structures testify to its longevity. Medieval builders were adept at employing the arch in their projects, as the Romanesque and Gothic cathedrals demonstrate. Arches served the same structural purposes in these churches that they did in bridges. Arches supporting the walkway, either singly or in sequence, rested upon piers on shore and often in the water. Building upon a solid foundation was often the greatest challenge for the medieval engineer. This was certainly true for foundations set into water. The Romans invented special cement to use underwater to hold the stone pilings together. Other techniques included the use of wooden pilings set deep into the bed of the river or stream. Once the wood was sunk into the riverbed, a temporary dam (called a cofferdam) was built around it so that water could be directed away from the piling and cement set on top of the wood. The resulting posts then formed part of the visible bridge structure. A fifteenth-century engineer described constructing a wooden form and then filling it with concrete to fabricate cement pilings. A traveler crossing an arch bridge was much higher above the water than one crossing either a beam or a pontoon bridge. Being made of stone, these bridges were much longer lasting, and many examples may still be seen today.

An example of an early stone bridge is the bridge spanning the Danube at Ratisbon, constructed over the years 1135–1146. The grandest bridge of medieval Europe was built over the Adda River, in northern Italy, at Trezzo. The bridge employed a single arch about 79 yards long and 23 yards high and took seven years to build, from 1370 to 1377. It was twice as long as the longest Roman bridge and would not be exceeded in length until the modern age (Garrison 1991, 109). Perhaps the most famous of the medieval stone arch bridges was London Bridge, which joined the shores of the Thames. Begun in 1176 and completed in 1209, it was more than 6 yards in total width (with a roadway of about 5 yards), about 350 yards long, and held up by 19 stone arches of various sizes. The bridge was so strong that shops and apartments sat on it. This was also true for medieval bridges in Paris and Florence (Boyer 1976, 75–76). The financing of London Bridge relied partially upon an increased wool tax. Paying it was the everyday contribution of most English people to this medieval technological marvel. Despite several fires, London Bridge remained standing until it was demolished in 1832 because the narrow arches and piers were affecting the flow of the Thames too much. That this bridge lasted more than 600 years is surely a testament to the engineering and technological know-how of medieval people. Recently, a scholar (Harrison 2004) has argued that the same number of bridges existed around 1750 as existed in the medieval era, which suggests a very sophisticated medieval transportation network.

CANALS AND WATER TRANSPORTATION

Long-distance transportation of people and merchandise rarely took place solely by either water or land. Any journey of substantial length would make use of both highways and waterways and employ the technology of both in pursuit of a final destination. The use of waterways for purposes of transportation stretches back far into antiquity. Scholars have demonstrated that water was the most cost-effective mode of transportation in the medieval period. All forms of rivers, streams, and artificial canals, as well as the sea, were used in the effort to move goods and people. Inland waterways were the preferred way to transport particularly heavy items. Stone and coal were common commodities shipped by water, rather than by carriage over land. The actual cost of stone was minimal; its expense came when the cost of transportation was factored in. With the explosion in cathedral construction, especially of Gothic variety,

The Vecchio Bridge in Florence, built in 1345.

inexpensive transportation of weighty stone became a pressing concern. Not only raw materials but finished products traveled via canals to coastal ports and from there to the rest of Europe. The ability to transport materials efficiently and cost effectively was crucial to the creation of manufacturing and the production of refined goods (Hutchinson 1994, 117, 119).

There were downsides to water transport. The most obvious of the limitations was that river transport was, in effect, a one-way street. It was remarkably easy to move cargo downstream, but traveling upstream required a great effort. Also, the same financial dangers posed by road transport also applied to waterways. There were tolls for passing over and under bridges that were owned by local lords. Should a barge or boat have an accident and be unable to pass under a bridge, the cargo belonged to the owner of the bridge (Leighton 1972, 95–96, 125).

Where a naturally occurring river or stream did not exist between points where commerce and thus transport needed to occur, medieval construction teams built canals and locks. Canals are an invention from antiquity that were used to deliver water where it did not occur naturally as an aid to transportation. Middle Eastern canals date from around 700 B.C.E. The Romans built canals mostly for military transport, but after the empire fell these languished and fell into disrepair and neglect. These Roman canals were reclaimed by medieval opportunists and eventually carried almost all commercial transports both human and cargo. The practice of creating new canals generally dates from the late medieval period. The Holy Roman Emperor Charlemagne attempted to create a canal off the Danube in an early, though unsuccessful, canal project. The growing economic markets characteristic of the later medieval era inspired new attempts at canal creation. This is an example of market forces driving technological innovation.

Problems of geography often stood in the way of building canals. Whereas a river carves its own path naturally out of the earth, artificial canals must rely on either major excavation projects or the use of locks. The Chinese invented the canal lock, and it was in regular use in late medieval Europe. A lock permits ships or barges to move from one water level to the next and allows the vessel even to travel up and over obstacles. Ships could travel uphill a little bit by going against the flow of a slightly downhill river or stream, but more drastic gains or loses in elevation required the use of a lock. The simplest form of a lock is called a "flash lock." In this technology, a temporary dam is erected downstream, which allows the water level to rise and the ship to sail upstream more easily. A more complicated type of lock involves a chamber of changing water levels. It functions as follows. A ship sails into a sort of holding pond between an upper and a lower level of water, and a gate closes behind the ship. Then water from the upper level is allowed to flow into the holding area, and the ship begins to rise with the water. When the water holding the ship is equal in height with the water at the upper level, the gate on the opposite side of the lock is opened, and ship continues on its journey. Ships wishing to sail in the opposite direction (to a lower level of water) follow the process in reverse.

TRANSPORT SHIPS AND DOCKS

The earliest boats for both people and cargo were rafts made of logs bound together. Between the fourth and ninth centuries, Irish and Scottish adventurers sailed in thinly framed boats covered in animals skins. When profit and plunder were at stake, advances in cargo capacity and ship size increased dramatically and rapidly. For example, ships dating from the era of the Crusades "could carry up to 600 tons of warriors, horses and supplies" (Whitney 2004, 123).

The majority of ocean-going vessels in medieval Europe were not military in purpose but rather dedicated merchant and transportation ships, although these could serve double duty by ferrying men and supplies. For example, invasion forces often crossed to their destinations on rented merchant ships. This was not always a smooth business transaction. Riots often resulted when Crusaders could not pay for their transportation to the Holy Land. Less glamorous transport ships included ferries and barges. Some inland transportation ships were very simple in construction, though nonetheless effective. Boats that date from the twelfth century and that consist of logs lashed together have been found in England, and certainly these would have been efficient barges. More sophisticated vessels were constructed out of planks. These had flat bottoms with triangular cross-sections and were designed specifically to hold the maximum amount of cargo (Hutchinson 1994, 122).

The coming of the Crusades in the eleventh century created a demand for ships with increased carrying capacity, an example of religion spurring new technology. Crusaders did not travel light but rather journeyed with vast amounts of military equipment, as well as great numbers of horses. The round ship (so called because of the shape of the hull) was most commonly used to carry cargo. These ships also employed a contingent of soldiers to fight off pirates should the need arise. Because of the dangers posed by open water, cargo ships rarely ventured too far from the coast. The ships moved slowly toward their destinations because they often had to wait for favorable winds. Later, Italian ships, built in anticipation of the Ninth Crusade, utilized the latest in contemporary technology. These new cargo ships had two masts, each carrying a triangular lateen sail. They steered by means of two side rudders (on port and starboard). Cargo was removed from the hold through a large opening in the main deck. On the deck were larger-than-normal castles to house nobility traveling east to participate in the Crusades. Merchant ships had much smaller versions of these enclosed living spaces at the stem and the stern. As the demand for larger cargoes increased, so too did the number of decks inside transport ships. The heaviest items found a home on the lowest deck, but animal and human cargo could now be increased in the same way that bunk beds increase the capacity in children's bedrooms. Specialized construction of the round ship, which lent itself so well to cargo carriage, relegated other former cargo ships such as carracks to functioning as warships (Unger 1980, 122–126).

When cargo ships arrived at their location, they had to be unloaded at a dock. The coast of northwest Europe was lined with port cities that served as bases of trade between England and the rest of the Continent. The earliest archaeological remains of a medieval dock in England date from the early thirteenth century. When it functioned, the dock was longer than 22 yards in length, with walls on either side made of stone. If a large ship could not get close enough to an inland port to unload its cargo, smaller ships served as go-betweens. Cranes were a common spectacle along harbors from at least the twelfth century. Their use on inland building projects (such as grand cathedrals) dates from around the same time and may well have been a parallel technological development. One type of harbor crane was the "hoisting spare," which consisted of a vertical post with a horizontal cross-member that could pivot up and down in addition to rotating the cargo out of the ship's hold onto the dock. The second style of crane was the "windlass," which operated by means of a large wheel that pulled, or released, a rope that traveled along the boom arm of the crane and pulled cargo up out of the vessel. Cargo was often so valuable and irreplaceable that captains of ships had to show merchants the quality of the rope that was going to be used to hoist the cargo out of the ship's hold before loading or unloading could commence.

OCEAN TRADE ROUTES

Although the most celebrated examples of medieval trade occurred over great distances, much of the trade conducted by the majority of people in their everyday lives took place on a local community level. As agricultural production increased and generated a surplus, weekly fairs allowed people to engage in a kind of barter trade, exchanging excess harvests for everyday staples such as cloth. Large annual fairs attracted international merchants who brought all sorts of luxury goods such as foreign spices to European markets. Even though most of the people in medieval Europe never ventured very far from their homes, they were able to purchase interesting and exotic goods from those who did.

While moving goods on inland waterways was relatively worry-free (save for the occasional tolls charged to ships sailing under bridges), ocean-going cargo ships faced a host of potential calamities. "For a land civilisation like that of the Middle Ages, the sea could only provoke fear, anxiety, and repulsion." As late as the thirteenth century, the ocean was thought to be home to all kinds of evil creatures. Sirens, with beautiful female faces, sat upon coastal rocks tempting sailors to steer toward their jagged perches. The ocean was also filled with a fury that appeared in unforgiving storms. Some fears were well founded. Conquerors came from the sea, and the sea brought disease from far-off lands. From the ninth century on, pirates in the Mediterranean threatened all ships that traveled in those waters. In the same way that highway robbers harassed travelers on roads, so, too, did pirates on the water. They would lie off coasts and block entry into ports. The English Channel was particularly dangerous (Verdon 2003, 63, 65). The best defense against pirates was to sail in convoys.

And yet, despite all these dangers, the open water was essential to the medieval economy. Technologies of shipbuilding and navigation that aided medieval exploration were the same ones that created international trade. Those who braved the mysterious waters had to balance their natural fear of the unknown against the potential of fabulous riches.

Not surprisingly, coastal towns and cities became renowned for their traders and merchants. The best example of this occurred in Italy, where cities such as Venice, which grew none of its own food, were keen traders. The Crusades brought Italians into contact with the riches of eastern countries and peoples. Practically every Italian ship that carried Crusaders to the Holy Land returned with bolts of silk and barrels of spices. European merchants also relied on contact with Arab traders who sailed toward the West carrying a variety of highly prized spices, including pepper, cinnamon, and nutmeg. Webs of trade routes centered at major European cities such as Constantinople were linked to Venice, Naples, and Genoa and ran through Baghdad, Alexandria, and Tripoli. Inland trade routes crossed much of continental Europe, specifically the Holy Roman Empire (modern Germany) and passed through cities such as Augsburg, Vienna, and Nuremburg. Ironically, the same trade routes that brought riches and wealth to much of Europe also brought the worst epidemic the world had seen. The Black Death (1347) traveled on returning trader ships first to Italy and then to the rest of Europe along established mercantile routes.

The constant warfare of the medieval era did not mean that enemy nations ceased their trading relationship. The Hanseatic League, which joined German towns and the Baltic coast into a trading power from the thirteenth century on, dominated northern European trade. The Baltic areas exported mostly grain, various metals, wood, pitch, fur, wax, and fish. Imports consisted of wool, cloth, some manufactured items, and wine. Although they would be at war for much the medieval period, England and France had an extensive trading relationship that required a great deal of shipping across the English Channel. Also important was the feudal relationship between French kings and their England vassals, who had to make several trips back to the home of their ancestors. Certain regions of France traded more extensively with England than did others, for example, Bordeaux, which was as famous for quality wine in the medieval period as it is today. Even during the Hundred Years War, trade between France and England continued.

England's chief export was raw wool. The twelfth-century Flemish cloth industry depended heavily on English wool, and a regular trade route was established. Italy, too, imported a great deal of wool from England and Scotland. Initially, the wool was shipped across the Channel and then carried overland into Italy. After a route through the Straits of Gibraltar was established toward the end of the thirteenth century, water transport superseded the former two-stage process of getting the wool to Italian cloth manufacturers. In exchange for wool, the Italians offered luxury items obtained from the East. Because Italian merchants tended to deal in small-volume cargo of extreme value, their ships (galleons) tended not to have the capacity of those of their northern European counterparts (Hutchinson 1994, 77, 84).

4

COMMUNICATION

In the modern world, communication assisted by technology is rarely given a second thought: picking up the phone or sending an email to people near and far is a common experience. Local, national, and international news is available at the push of a button on a television or computer. Newspapers arrive on the doorstep in the middle of the night. Twenty-first-century life is so interconnected through communication and real-time reporting of worldly affairs that people have to decide whether they wish not to participate rather than to solicit this information. In the medieval period, things were remarkably different.

This chapter reveals the various forms and locations of medieval communication. Things that today are taken for granted were major innovations during this period. For example, books were expensive handwritten items and something of a novelty. Paper, one piece of communications technology that is now everywhere, did not exist for much of the medieval period; it was first produced in Europe only in the thirteenth century. The developments of books and paper are treated in separate sections. Libraries grew from books stored on shelves on monastic walls to the great book depositories characteristic of late medieval universities. News might take weeks or months to travel, and, with no recognizable mail service, sending and receiving letters often meant waiting until a traveler happened to be going to the same destination as the letter. Some of the most beautiful medieval art is found on illuminated manuscripts, whose production is discussed here. The late years of the medieval era witnessed the invention and widespread use of printing. This had profound effects on book production and the dissemination of knowledge. Finally, poor or diminishing eyesight kept some people from participating in medieval communication until the invention of eyeglasses in the late thirteenth century permitted scholars to read and write well into their advanced years.

PAPER MAKING

The earliest paper in medieval Europe was not really paper at all. It was parchment, which was produced from prepared animal skin. This parchment was also called vellum, although technically vellum comes only from the skin of cows. Turning flesh into writing paper was the job of parchment makers, a profession that had existed since the ninth century. The process began with the selection of skin. France and England used the skin of sheep and cows, while goatskin was commonly used in Italy. There are instances of parchment coming from other animals, too, such as horses and even rabbits. Skin from unborn calves was particularly prized but hardly ever used because of the complicated process of obtaining it and the exorbitant price of this exotic writing material. Parchment makers had their own method of simply eyeing the animal whose skin that would yield the best parchment or vellum, in the case of cattle. After the skin had been removed, it had to be washed and the hair or wool separated. This often meant leaving the cleansed flesh out in the sun and waiting for the hair to rot off the hide. The still-wet skin (it had to be kept moist to prevent tearing) was then fixed to a frame and scraped to remove any remaining hair. (The nonhair side was the preferred writing side.) The skin was washed a second time using a process that lasted for two days. Once cleaned, the skin was stretched in a wooden frame and allowed to dry. When the parchment was dried and tight, it was scraped once more to achieve the desired thickness. The newly finished parchment was then ready to be sold (De Hamel 1992, 8, 11). Such a lengthy and laborious process meant that parchment was a costly and valuable commodity.

A less expensive and more easily made alternative to parchment appeared in Europe around the tenth or eleventh centuries C.E. This innovation was what modern eyes would recognize as paper, and it was made from a pulp mixture of linen rags and water. The process of making paper from pulp likely dates to 100 B.C.E., and the technique spread rather slowly throughout Europe. Almost certainly, China had paper (rather than animal-skin vellum) before any of its neighbors and long before Europe. Chinese paper was fashioned from the pulp of hemp, mulberry, and bamboo. From China the knowledge of making paper spread to the rest of Asia: Korea had paper by the sixth century, then Japan, followed by Tibet. Arab paper factories opened in Baghdad toward the end of the eighth century C.E. At Baghdad, the pulp was pounded by trip hammers operated by waterwheels. In the thirteenth century, similar paper mills appeared in

Spain and in Italy. France's first mechanical paper mill came on line in the fourteenth century. Despite this mechanized production, paper was still not widespread until the fifteenth century, when the invention of printing created a demand for vast quantities of inexpensive paper. For purposes of comparison, consider that a medieval book containing around 300 sheets of parchment required the skin of nearly 200 calves as opposed to a pile of cloth rag.

While later technological advances would speed the operation greatly, the procedure for making paper from linen or cloth remained basically unchanged for centuries. Linen or cotton cloth was cleaned and allowed to soak in a container of water for several days. The cloth was then cut into pieces and pounded for hours. At first this was done entirely by hand, but later on it was done by trip-hammers powered by waterwheels. The pulp was set aside and then beaten again. Once the mixture was very fine, a square mesh frame was used to pull out some of the pulp. On the frame there was now a very wet sheet of paper pulp. The sheet was carefully turned out of the mesh and placed on a piece of felt cloth the same size as the paper. A second piece of felt was laid on the top and another wet sheet of paper placed on the top of the felt. The process was repeated a number of times. A stack of alternating paper and felt was then placed in a press and the remaining water squeezed out. The individual pieces of paper were then allowed to dry. At first Christian scribes and authors were very suspicious of this kind of paper and, viewing it as a Muslim innovation, rejected it. They preferred the traditional (seen as somehow more Christian) medium of vellum. This aversion soon passed as the number of published works and demand from the new universities exceeded the capacity of the parchment makers, although important documents such as governmental and papal decrees continued to be written on parchment rather than paper (De Hamel 1992, 17). Despite the invention of paper, wax tablets continued to function as they had since classical antiquity, recording material that was not meant to be permanent. These devices were formed from a shallow wooden box frame into which wax was melted. Notes and messages could be etched into the hardened wax with a sharp pencil-like implement and then the tablet could be wiped clean and used again.

ILLUMINATED MANUSCRIPTS

While the vast majority of medieval manuscripts and books are simply black text on white pages, there are celebrated examples of decorated or illuminated manuscripts. In strictest terms, illuminated manuscripts are only those adorned with gold or silver leaf. The light hits the thin pieces of metal and "illuminates" the text. All other pages that are decorated with color are not technically illuminated. Nonetheless, for the purposes of this discussion, colored embellishment on medieval pages is referred to as illumination.

Illuminated manuscript page. The Pierpont Morgan
Library, New York. MS M. 183 f. 13r.

Medieval books had no title page in the manner of modern books. From
the early seventh century, a practice developed of making the first letter of
a work much larger and more elaborate than the others as a means to in-
troduce the book and the various chapter divisions within the body of the
text. Over time, the decorations also became more colorful. Making the initial letter of a section larger than the others was part of the process known
as the "hierarchy of decoration." This technique involved more than simple artistic consideration. As the historian Christopher De Hamel puts it,
the practice "is appropriate … for the Middle Ages when people had a
strong sense of the gradation of things. Angels, stars, animals, kingdoms
… and so on, were all classified into levels of unalterable rank and status
with much more assurance than today, and a similar hierarchical sense of
the order of things is inherent in the ornamentation of texts" (De Hamel

1992, 45). The size and detail of the illustrations varied greatly and ranged from adding one color to an entire page of pictorial splendor. Sometimes the words occupied fewer than four lines in the center of the page, which itself was more artwork than text. But, the pictures themselves were communication. Sacramental lessons, legends, and political events could all be artistically represented on the page. This was especially true in the books of hours (devotional manuals based on scriptural teachings), where the pictures were often more studied than the actual words (Griffiths and Pearsall 1989, 38). Indeed, the majority of manuscripts produced by monastic scribes were not original compositions but rather copies of important texts or devotional manuals. Some monastic orders were prohibited from using metallic decoration because it was thought to be garish for a devotional project. Where it was permitted, gold or silver for manuscripts came in two forms: leaf to be attached by glue or powered gold to be mixed with an appropriate adhesive.

The illumination would be added to the page after the text had been set down. This meant that the space reserved for the pictures had to be roughed in before the scribe could write the textual portion of the page. This procedure is known from the existence of unfinished manuscript pages dating from the medieval period. Scribes wrote on sheets with pre-existing lines or lightly drawn lines before proceeding. Prior to the twelfth century, a knife blade or a similar tool scored guidelines on the page. After this time, lines were set by pencils and later by colored ink. The two most common colors in medieval manuscripts (aside from black) were red and blue. Producing black ink involved mixing charcoal with gum, while red ink came from combining either powdered mercury or vinegar-soaked brazilwood chips with gum. Blue ink was created from smashed rocks that had a bluish tinge. Because of the addition of gum (usually a viscous plant sap), ink used in medieval manuscripts was much thicker than modern versions. Scribes and illuminators deposited ink on the page with a quill pen made from a feather or piece of thin reed. The choicest feathers came from geese; these were soaked and then heated to make them hard yet flexible. Once the pages of the book were complete, the volume was given to the stationer or, in the case of secular books, to the bookseller (religious books, too, in the late medieval period). The earliest monastic books were finished completely in house.

BOOK MAKING

For medieval readers what would be recognized as a modern book was known as a codex. This form of book differs markedly from that produced by the ancient Romans, who had written on papyrus fashioned into rows and then coiled on a scroll with text on only one side. As opposed to these scrolls and individual single-sheet manuscripts, a codex was characterized by hand-written pages bound between two covers. The process of making a codex was lengthy and involved some of the most skilled artisans of the medieval period.

In the words of P.J.M. Marks, curator of the subject at the British Library, "Bookbinding is the process whereby the pages ... are secured in a particular order and encased by protective covers" (Marks 1998, 9). The codex first appeared around the first century C.E. and was a Christian invention. Prior to this innovation of encasing pages between covers and sewing the entire volume at the spine, scrolls and tablets were the only method of "binding" available. Early European bookbinding was done by monks for their fellow brethren. Reading and learning were sacred monastic activities, so it comes as no surprise that bookbinding began with monastic volumes. Of course, not every medieval book was spiritual or devotional, and secular binders were responsible for legal and educational books.

The number of folds (and subsequent cuts) in a standard piece of parchment or paper determined the size of the book: a single fold is a folio, two folds yield a quarto, and three produce an octavo-size volume. Where the codex is made from sheets of parchment, hair side faces hair side and flesh side faces flesh side. Once the pieces were gathered into bundles of several sheets and arranged in the correct order, the next step was to sew the pages into what is known as a signature. A codex was made up of several signatures. The earliest method for securing the signatures to the covers was known as "stabbing"; the signatures were pierced through their spines, using existing holes that had been created, to secure the pages in the signature. This process of sewing is still used for high-quality hardcover books. Coptic bindings employed a type of chain stitching where part of each stitch entered the signature and then was joined to the next one on the outside. Most medieval books were formed using techniques of flexible sewing. In this process, the signatures were sewn around horizontal bands (cords) set on the spine. The results may be seen on any number of older books in libraries where the bands are visible as horizontal bumps along the cover spine of the completed book. Later books were made using a similar technique that relied on recessed cords that gave the finished book a smooth spine. To protect the first page of text from the cover boards, medieval bookbinders glued end leaves on the inside of the cover. The other side of the end leaf was left free to form another page, called a flyleaf, in the book. To prevent the weight of the pages from pulling the spine downward, binders were used to reinforce the spine so that it would maintain its shape.

The earliest medieval books were boarded with heavy wooden covers. These served to protect the parchment and to ensure that it did not curl. Covering the wood with leather was a later innovation. The thickness of the wood (usually oak) with the leather outer cover ranged from 0.25 to 0.65 inches. In addition, some binders attached clasps on the open end of the book so that it could be fastened to ensure that the parchment remained flat. Some treasured religious works were covered in gold encrusted with gems. However, strictly speaking, these are more valuable dust covers than proper bindings. In the sixteenth century, pasteboard (sheets of paper glued together) began to replace wood as a rigid cover material. The advent of paper meant that pages were no longer subject to curling and that heavy covers were therefore no longer needed. The increasing prevalence of paper meant that books could be made much more cheaply, and the demand for them grew. This new market for quickly produced volumes initiated the creation of the paper-bound book, or the modern paperback. Paperbound books did not wear particularly well, and the paper covers were often used only until the owner could get the book bound in hard covers (Szirmai 1999, 103, 107, 132). As they do today, dust jackets covered books. These covers were of woven cloth and, in the case of devotional or religious works, embroidered with figures from scripture. Because of their delicate nature, very few of these have survived.

Islamic bookbinding differed from that practiced in Christian Europe. Rather than employing thick bands to secure a stitched binding, bookbinders in Islamic regions glued the signatures of their books to the covers, which themselves were very thin. Most Islamic book covers included an extra piece of leather that wrapped around from the back over the text edge and folded over about half of the front cover. The exact purpose of this covering is not known, but it likely served to protect the pages (Avrin 1991, 312).

Medieval book storage was different from the modern practice of shelving books vertically with the spine facing outward, and thus different considerations were at play when the binder neared the final stages of binding. Until the sixteenth century, books were shelved horizontally with the open edge facing outward. This required that the binder pay more attention to the appearance of that side of the book. Often the name of the author and the title of the book were written on this edge. Other binders decorated the open edge with colors and designs. Around the 1460s, first in Italy but soon in other areas, as well, the practice of gilding the edges with a fine layer of gold developed (Marks 1998, 49–50).

The text of the books was entirely copied by hand until the middle of the fifteenth century and the arrival of printing. Even after this date, many books were still copied by scribes. The process was a long one. Monastic copyists could expect to produce three or four books a year because they could not devote themselves fully to the task. Alternatively, professional writers, being paid by the book, could be expected to complete a new book every few days. (Of course, this estimation depended on the length of the codex.) Early in the fifteenth century, people who desired specific titles simply went to the local bookshop and ordered a copy. Books were still mostly one-of-a-kind products and all made by the hands of craftsmen. These people should not be confused with authors who composed the original content. Until well past the medieval period, most books were not sold in prebound states. Only upon orders were the loose sheets making up the contents of a particular book bound for the customer. However, some books, such as devotional works, those by certain classical authors, and school texts, were so popular that booksellers carried a number of ready bound copies in stock.

Universities created a demand for more books and standardized copies of key textbooks. This led to a new system of production, known as the *pecia* system. All textbooks were first checked by faculty members and an authoritative copy agreed upon. The signatures making up this prototype, called a *pecia*, were then distributed among different scribes in a kind of assembly-line production of as many copies of the book as were needed for a particular class. At Bologna in the thirteenth and fourteenth centuries, this procedure was used frequently. For students who could not afford a personal copy (approximately fourteen shillings each in 1479 or about $600 in modern dollars), booksellers would rent out the finished books for the duration of the semester at a rate of one pence ($3.00 in the modern equivalent) for every sixty lines of text. Another key development was the introduction of spaces between words in the later medieval books; in the early medieval era, words simply ran together. Modern punctuation appeared only around 1500.

LIBRARIES

Libraries existed in Egypt and Mesopotamia as early as 4000 B.C.E. The "books" housed in these earliest libraries were papyrus scrolls and clay tablets. The greatest of these ancient libraries was a product of Greek conquest, by Alexander the Great, in the founding of Alexandria, Egypt. The celebrated library at Alexandria was more than a storage facility for the known books (scrolls) of the time; it was the center of learned activity for hundreds of years. At Alexandria, scholars copied, collated, and edited much of the knowledge then extant in the world. Every book known to exist within the city was copied and the duplicate placed in the library. Any ship that arrived in the port of Alexandria was required by law to lend to the library any books carried on it so that copies might be made. At its peak, the library was believed to hold more than 600,000 scrolls on as many different subjects as one could imagine. The golden days of learning and literature spawned by the library did not survive the violent civil wars in Egypt and invading Roman emperors. Between 89 B.C.E. and 273 C.E., parts of the city and the library were burned to the ground. A final conflagration consumed the remains of the library in 645.

The constant invasions and societal upheaval that characterized the early medieval period following the collapse of the Roman Empire in western Europe meant that the grand libraries of antiquity continued to exist only in the remaining Eastern Empire (called the Byzantine Empire). During this time, Arabic scholars surpassed their Christian counterparts in terms of libraries. By the tenth century, the more than 30 public libraries, founded by the Abbasid dynasty, that existed in Baghdad, the center of Muslim learning and culture, rivaled any other library in the world in size and content (Battles 2003, 22–25). After the Roman Empire was divided and the Emperor Constantine settled in Byzantium, renaming it Constantinople, he established an imperial library in the city that bore his name. This library served as an eastern center of learning from its opening in the fourth century c.e. until the Turks sacked Constantinople in 1453. It was from the holdings in Constantine's library that western scholars rediscovered the forgotten riches of Greek philosophy such as the writings of Aristotle, Plato, and Galen. There is a certain amount of irony, then, that the greatest damage done to the library was committed by Christian crusaders in 1204 who viewed the books and manuscripts as commodities to be looted and sold. This view of books was not uncommon. For example, Charlemagne saw books as the spoils of war and often seized volumes from those he conquered. His library is known to have been one of the best in all of early medieval Europe. As the historian Bernhard Bischoff has put it, the goal of Charlemagne's extensive library was to "preserve the literary treasures of the past." Upon his death, the numerous books were sold to raise money for the poor (Bischoff 1994, 56, 62, 94).

In western Europe, libraries in monasteries were kept within chests and on small shelves. Monasteries greatly valued their books, made copies of them, and used a network of monks to obtain more. Under the direction provide by the Rule of St. Benedict (ca. 480–543), reading and copying of devotional works became part of the monks' everyday tasks. Cathedral schools, where both future clergy and younger secular students received an education, had more diversity in the contents of their libraries than did monasteries. The demands of learning required that these libraries broaden their collections to include both devotional and philosophical titles. Libraries at cathedral schools were generally much larger than monastic libraries, which tended to be satisfied with holding one book for every monk. After this ratio had been achieved, the growth of monastic libraries tended to stagnate. The core collections of monastic and cathedral libraries were similar. Of greatest importance in both libraries was the Bible (several copies), followed by works by patristic authors, devotional manuals and books of hours, and, last, books of Latin grammar.

By the 1300s, monastic and cathedral libraries had increased in size to include books by many classical Latin authors. This increase in volumes required a system of organization, usually into secular and nonsecular titles. By the fifteenth century, book chests had long since been replaced by small rooms and, eventually, separate sections in monasteries or cathedrals set off as dedicated libraries. Often, parts of the libraries were public. With more users came the possibility of theft. This danger was combated with the practice of chaining books to their shelves. At first only the most popular books, rather than the most valuable, were chained, but eventually entire libraries were secured to the shelves by chains. Books were simply too valuable to be unsecured. Universities, which began as a collection of students around a master, did not at first possess libraries. Only in the thirteenth century, with donations of books from patrons such as Robert de Sorbonne and Robert Grosseteste, did universities like the ones in Paris and Oxford begin to acquire their collections of books. In the fourteenth century, plans were made for a dedicated library building at the University of Oxford, but designation of a specific location had to wait until after the generous donation by Humphrey, Duke of Glouces- ter (brother to Henry V), of many theological books. Oxford acquired its centralized library, known today as the Bodleian, in 1602, and scholars may still visit the Duke Humphrey Reading Room (Harris 1995, 100, 102, 110).

DISTRIBUTING NEWS AND DELIVERING LETTERS

Delivering news and letters involved many of the technologies described in this book. For example, messengers traveled by horseback or in carts on roads, crossed waterways on bridges, and sailed on ships in the ocean or on inland canals. People of all classes relied on commu- nication technologies and methods. For peasants and persons of lesser social standing, much of their experience with news and letters was in the passive role of reception of messages such as those delivered by wandering preachers. Many clergy saw this form of distributing God's good news as a pious duty. Since most people lived within a day or so's journey from the rest of their family and were either illiterate or in possession of only rudimentary literary skills, communication was oral and the spread of information the result of personal travel (McKitterick 2001, 23, 227–228).

For news of national or political importance and for business transac- tions, a more complicated system was used. During the chaotic period following the fall of the Roman Empire, which had possessed an impres- sive system of communication, Catholic clergy were often used as mes- sengers and to carry important documents. There were two reasons for this choice. First, Catholic clergy were literate in an age when the ability to read was crumbling as fast as the marble remains of the once great Empire. Second, priests were generally left alone as they traveled. Clergy were not, however, always immune to searches. Important letters had to be hidden inside other objects, such as beneath the wax of a writing tablet. Latin was the language of choice for high-level communications. It was still used in scholarship and government but could not be read by the majority of people even if they could read their vernacular lan- guage. This was an important consideration should a sensitive message be intercepted on route. The Catholic Church itself needed a secure and more permanent method to distribute decrees and doctrine. As a result, the early Church undertook road repairs and bridge construction and established a series of inns so that weary messengers could have a safe haven as they traveled to deliver the ecclesiastical news. Relays of messengers ensured that no one person had to traverse the entire distance that a letter had to cover.

When lay couriers were used, the messages they carried were often oral. Information was contained within the memory of these lower-status messengers. Similarly, letter carriers tended to be people who happened to be going to where someone wanted a letter sent. These people had the advantage of natural camouflage because they simply blended into their surroundings. In England, for example, around the middle of the tenth century, the Bishop of Worcester employed a group of noblemen who delivered episcopal messages by horseback. Nearer the time of the Norman Conquest (1066), men of lesser standing too are known to have carried messages from bishops (Leighton 1972, 19–21). Those carrying letters from kings and other crown officials had to swear an oath of secrecy. Secular rulers tended to rely on oral communication, which was more effective at enforcing law and order on feudal estates than was a posted document. In the case of very important news, such as the death of a monarch, a messenger would wander the countryside reading the news from an official account. As the historian Sophia Menache has noted, "medieval communication was characterized by the immediate contact between the communicator and his audience" (Menache 1990, 9).

Letters and news traveled only as fast as the messenger who delivered them did. For a person walking on foot, the best that could be achieved was about 20 miles a day. During peacetime, messages and news took days, if not months, to travel across the continent. Should the situation be dire, such as during war, then the time lag between event and reporting could be substantially shorter, though never less than several days or a week. Delivering news and letters was often so important that royal messengers in England had the royal right of purveyance, which granted them the power to demand use of any horse or cart to aid them in their duties. Secular and ecclesiastical rulers alike relied on news reports from people who traveled as part of their way of life. Hence, merchants and itinerant preachers, to name only two such groups, would often find themselves being offered rewards for any news that they might bring with them (Menache 1990, 11–12; Beale 1998, 34). More predictable sources of news were also used during the medieval period. The papacy employed "papal legates" to act as delegates and messengers to carry the orders and news of the Church to sacred and secular leaders in Europe. Kings used *nuncii* to carry political messages. Towns often employed their own couriers to dispense information of a commercial nature. France had a *poste royale* from the fifteenth century on. The Holy Roman

Empire (modern Germany) established a postal service beginning in the thirteenth century.

PRINTING

Printing with wooden blocks had been used prior to the invention of the printing press by Johannes Gutenberg in the middle of the fifteenth century. China, for example, had printing technology long before Europe. Although Gutenberg gets the credit, at least four other people attempted to create a printing press that used moveable type at the same time. Early printed items in medieval Europe included playing cards and broadsides (posters). The majority of printed broadsides were produced in what today is the south of Germany and contained religious images similar to those on stained-glass windows. The process to make the print began with carving the image into a block of wood; hence the term *woodcuts* to characterize medieval and early-modern illustrations. The block was then inked and the page pressed against it. Any coloring to the image was done by hand. Woodcuts were also frequently employed to print the large initial letters and illustrations in a medieval book, known then as a codex. Wooden blocks, however, can print only one thing and tended to wear out very quickly with repeated use.

Johannes Gutenberg (1397–1468) was born in the city of Mainz. His early career was as a goldsmith, and in March 1434 Gutenberg joined the guild of goldsmiths in Strasbourg. While in that city, he began construction of a new form of printing using individual letters. Gutenberg also experimented with a modified winepress that would be used for printing. Between 1450 and 1452, now back in Mainz, Gutenberg borrowed 1,600 gold guilders (approximately $1 million in modern currency) from one Johann Fust, a lawyer from a family of wealthy merchants, to finance his effort to build a printing press (Kilgour 1998, 85). In 1455, Gutenberg began selling a two-volume printed version of the Bible. Although modern scholars often celebrate the printing press for lowering the price of books and allowing knowledge to become more widely distributed, this was not the case with the first printed book. The cost of this Bible was tremendous, somewhere in the ballpark of three years' wages (about $1,650 in modern money) for an average worker. Even so, this was substantially cheaper than a handwritten version. This Bible, known as the Gutenberg or 42-line Bible (the number of text lines per page), ran nearly 1,300 pages in two volumes. The number of copies made was around 180, 40 of which were on parchment and 140 on paper. While this seems a small edition by modern standards, for the medieval period it was substantial. Even though the text was printed, the illumination and other decorations were entirely drawn by hand. As a result, each Gutenberg Bible is unique. (The few remaining Gutenberg Bibles in the modern world retain their great value. Complete books fetch nearly $30 million at auction, and separated pages may cost as much as $60,000 each.) Printing and finishing the 180 Bibles, using 6 presses each producing 16 pages an hour, took three years, or about the time required to make one hand-copied Bible. The profit from this initial batch of Bibles was not enough for Gutenberg to repay the loan from Fust, who then sued and was awarded possession of the press and the remaining Bibles. A bankrupt Gutenberg spent his remaining days operating a print shop in Mainz.

For his Bible, Gutenberg manufactured not only both the lowercase and capital letters of the Latin alphabet but also a variety of other characters, nearly 300, so that his printed Bible would look like a handwritten manuscript version. The job of making the typeface for his printing press was the most time-consuming part of the process, because a master-copy of every letter was required. Using his goldsmith skills, Gutenberg first carved a mirror image of each letter into the end of a metal bar to make a stamp. He then hammered the image on the stamp into a softer metal bar made of copper. The impression of the letter in the copper created a mold with which Gutenberg could make as many identical copies of the same letter as he wished. Doing so was the next step of the process. Molten lead alloy was used to cast the letters for the printing press. Once hundreds of letters of various types were ready, the task of printing turned to setting the text of the page into the press. After the correct order of the letters was ensured and the spelling verified, the letters were inked using special leather pads called inkballs and paper or vellum placed over the top. The press was brought down over the page and tightened. The first sheet was a test and was carefully scrutinized for errors. A single inking could be used to print hundreds of pages (Geck 1968, 26, 33–34). This new method of printing required a new kind of ink. The water-based inks used in block printing were unsuitable for metal type because water does not stick to metal. Gutenberg either invented or adapted (scholars are divided on this point) an oil-based ink, much like that used in paintings, that adhered to the metal letters.

The technology of the printing press spread very quickly throughout Europe. By 1480, more than 110 towns had presses, and by the turn of the century that number had risen to 326 towns, which had printed more than 35,000 titles. Further developments in the nature of books occurred at the same time that printing was taking Europe by storm. Title pages and page numbers both appear in books dating from around 1500. While page numbers may seem a second-rate technology, they allowed for indices, which in turn allowed for the easier systematization of knowledge (Kilgour 1998, 93).

The impact of printing on medieval society remains a debated topic among historians. Typified by Elizabeth L. Eisenstein, some scholars argue passionately that the arrival of printing was revolutionary and forever changed Europe. For Eisenstein, the transformation from "scribal culture" to "print culture" was earth shaking and influenced all aspects of society. She maintains that the technology of the press itself was the driving factor (an "intrinsic power") that moved Europe to knowledge preserved by print rather than by manuscript; in this way knowledge became permanent. More recently, many historians, most prominently Adrian Johns, have argued for an evolutionary interpretation of printing's impact and have rejected what is viewed as Eisentein's technological determinism. They note that printed books simply continued practices already common among handwritten books, such as tables of contents, mass production, standardization of contents, and the like. What is more, they emphasize that handwritten books were still produced for decades after the arrival of printing and that printed books mimicked handwritten ones in order to be more accepted by a society that had previously experienced only manuscripts (Eisenstein 1983; Johns 1998). The debate seems unlikely to be settled any time soon.

READING GLASSES

Prior to the invention of eyeglasses, aged scholars were forced to read through glasses of water or to have younger assistants read to them should they wish to continue their academic pursuits into their years of diminishing vision. Although lenses of various strengths had been known and ground since antiquity, they were employed mostly for burning and magnification purposes and not as an aid to eyesight. The English philosopher Roger Bacon, writing in the thirteenth century, noted that spherical lenses ground so that they were concave would allow readers with "weak eyes" to "see the letters far better." In 1289, an Italian commented that "I am so debilitated by age that without the glasses known as spectacles, I would no longer be able to read or write. These have recently been invented for the benefit of poor old people whose sight has become weakened." Thus, eyeglasses were a common item by the end of the thirteenth century and likely a few decades earlier. Dominican friars were perhaps the earliest users of eyeglasses. Learning and reading were at the heart of their mission, so it is natural that they would embrace any device that enabled them to continue reading even into old age. While the actual creator of eyeglasses remains unknown to modern scholars, the reputed inventor was an Italian named Salvino degli Armati (d. 1317), although there is no corroborating evidence. From about 1300, eyeglass manufacturing was common in Venice, a center renowned for glass production of all types. Eyeglasses proved tremendously popular. England, for example, imported nearly 400 pairs a month from the Continent during 1384, and within a century the number was nearing 500 pairs a month (Kilgour 1998, 77).

The knowledge of how to make eyeglasses (grinding glass lenses with a specific curve and then polishing them) was a closely guarded secret among the artisans who made them. The use of lenses to correct faulty vision may be traced to crossover technology from scientific investigation with magnifying glasses. In both cases, the lens extended the natural capacity of vision. The earliest eyeglasses worn in the medieval period did not have arms that wrapped around the ears. Rather, the circular lenses were surrounded by a metal frame and the two lenses joined by an inverted "V" which the wearer held at the apex to secure the lenses over the eyes. Many users preferred a single lens held in front of one eye, much like a modern magnifying glass. Variations included types of monocles. Because these spectacles had no arms, they were not worn constantly but only when one was reading or when close observation was required. Some innovative eyeglass users fixed their spectacles to the front of hats. This modification allowed the wearer to have both hands free to perform tasks other than holding the lenses over the eyes. Others, less concerned with fashion, attached a metal band between the lenses and rested the band on the bridge of their nose and their forehead until it sat on the top of their head. Earpieces were a later adaptation, first being made of leather (think 1920s motorcar goggles) and much later of a rigid material that hooked around the ears.

The exact function performed by eyeglasses was the subject of much learned discussion in the medieval period. It was known that they worked, but not exactly how. Some speculated that eyeglasses concentrated the image so that it might be more easily seen, while others suggested that the lenses aided in pupil control, which was thought to diminish with age. Another group noted that imperfect vision resulted from a defective curve on the lens of the eye for which eyeglasses compensated. This was an important consideration because philosophers believed, on the basis of the writings of Galen and of Arabic authors, that the lens of the eye refracted light and projected it to the interior of the eye, where it inter-acted with the crystalline lens, which was the organ of sight (Wade 1998, 49–50).

Very quickly after the invention of eyeglasses, the wearing of such aids to vision acquired for their user the modern connotation of book-ishness, learning, and erudition. Fourteenth-century paintings of famed monks and saints depict them using eyeglasses even when the subjects' lives predated the invention. The inclusion of eyeglasses on a portrait of St. Peter, for example, was likely intended to convey the saint's honor and reputation for learning (Corson 1967, 19, 21, 23).

5

EXPLORATION AT SEA
AND ON LAND

When Christopher Columbus sailed in 1492 with the goal of establishing a trade route to Asia, he did so with an understanding of geography and navigation that can be traced to the medieval era. What is more, the ships in which he made the famed journey were also of medieval origin. Columbus was the inheritor of centuries of cartography, travel literature, and technological developments that are the subject of this chapter. For example, when Columbus reached the shores of Central America, which he took to be Asia, he was surprised to see that the region was nothing like the description in Marco Polo's account from the thirteenth century. Moreover, the monsters that were believed to inhabit far-off lands were nowhere to be seen. The expectation that they might be encountered reveals how steeped Columbus was in the medieval mindset of foreign lands (Phillips and Phillips 1992).

Because knowing where you are is the first step in exploration, this chapter begins with separate accounts of medieval geography and cartography. Both of these areas of study were conducted within a theological framework that directed how the world would be depicted. One cannot go anywhere without proper transportation, and thus shipbuilding is considered next. Medieval ships were near the pinnacle of technological advance. Other emerging technologies such as the compass and the rudder are examined, too. Without sails, ships literally stood still, and the various shapes and materials are described here. Reports and rumors of faraway lands added an air of mystery to travel and exploration and often combined with a sense of Christian duty to explore the world; the chapter concludes with some of these myths and fables.

THEORETICAL GEOGRAPHY

Modern descriptions of medieval studies are not always representative of the actual practice. For example, there was no consistent medieval word that corresponds directly with our modern term "geography" with its connotation of dispassionate scientific study of the earth and its physical features. Only rarely did writers employ the Greek word *geographica* from which the term "geography" comes. The modern term did become more common around the fifteenth century. Prior to this, however, there was no standard designation for the study of the earth as a separate discipline. Sometimes another Greek term, *cosmographia* (pertaining to the study of the earth as a whole, including the earth as a planet), appeared in the literature but with no consistency of usage. Contemplation of the earth took place within a Christian framework; it was never a secular subject. One of the Church fathers, St. Augustine, encouraged examination of the world because he believed that Christians should be at least as knowledgeable as non-Christian scholars, such as ancient Greek and Roman writers. For Augustine, geographical knowledge was a part of a larger set of knowledge called *scientia* (the knowledge of human things). *Scientia*, in turn, was meant to support the truly important knowledge of *sapientia* (knowledge of divine things).

Sapientia and *scientia* merged in Bible studies. Monks, for example, were to learn some geography because it provided them with the proper understanding of the locations described in scripture, the most important of which was Jerusalem in the Holy Land. As the historian Natalia Lozovsky puts it, "The wish to experience the Holy Land—to walk through it, to see and probably touch the holy places mentioned in the Bible—was inspiring pilgrims throughout the Middle Ages." Not everyone would make the journey, and for those who did not, accurate descriptions of "every location, every physical detail of the landscape" would be their only encounter with the lands of the Bible. Thus, works of geography created a kind of virtual spiritual experience (Lozovsky 2000, 46).

According to the historian Elspeth Whitney, medieval studies of geography were conducted as a subset of astronomy. Topics that fell under the subject matter of geography included proving that the earth was a sphere, determining the size of the earth and of the three known continents (Europe, Asia, and Africa), and pinpointing their locations. More speculative elements included pondering the likelihood of an unknown fourth continent (Whitney 2004, 107). Like many other scientific endeavors of the medieval period, geography was based upon earlier Roman writings such as Pliny the Elder's *Natural History*. Classical authors provided medieval thinkers with methods and the precedent of relying upon written authority when composing geographical descriptions: something was true because an acknowledged authority had claimed it to be true, not because of personal investigation. A major influence in the study and teaching of medieval geography was the work of Isidore of Seville (ca. 570–636), especially his *Etymologies,* an encyclopedia that focused on the etymology (history of words) of place names. While this might seem unusual to modern readers, it was serious medieval scholarship. Names were more than ways to differentiate things; they held the key to the essence and being of all things. Within the names of places and geographical features were their origins and purposes. The world was God's creation, and everything held an imprint of divine purpose. Thus, the end goal of medieval geography was very different from the modern one. Again, the historian Natalia Lozovsky: "Useless in everyday practice, theoretical geographical knowledge served entirely different purposes, mainly defined by the ultimate goal of medieval studies of the natural world—to understand God by means of understanding his creation" (Lozovsky 2000, 153).

In the seventh century, Muslims traveled from Arabia to spread the Islamic religion and the teaching of the prophet Muhammad. By around 732 c.e., they controlled much of what is now the Middle East, as well as parts of North Africa, Spain, and Portugal. This enormous empire led to great interest in geography and cartography on the part of Muslim scholars who wished to accurately describe their empire. Continued contact with the Byzantine Empire (formerly the eastern half of the Roman Empire) meant that Arabs had access to many learned works from antiquity that would not be recovered in Christian Europe for centuries. Thus, Islamic geographic knowledge was for a long time superior to that held by Christians. For example, Avicenna, the philosopher and commentator on Aristotle, argued that mountain valleys resulted from the lengthy process of rock erosion caused by rivers (Martin 2005, 47–49).

MAPS

There were two basic types of geographic studies. The first type was theoretical, and the second variety was more practical but still possessed a strong theoretical or idealized component. Theoretical geography considered the world as a whole and as such was of little or no interest to the everyday traveler. Geographical knowledge possessed by the majority of Europeans amounted to practical things such as where the shallow part of the river was, which paths were safe, which slope was unstable, and other everyday concerns. Drawing and producing maps (now part of the modern discipline of cartography) were part of the second type of geography, although theoretical and historical maps did exist that were based on scriptural accounts of biblical times.

The earliest medieval maps addressed the known world in large scale. Commonly, maps made prior to the twelfth and thirteenth centuries depicted the world in its entirety or the land masses believed to be inhabited. After this time, regional maps appeared, and in the fourteenth century the first maritime maps were used. Ancient influence is evident in medieval maps. Greek philosophers saw the earth as a giant globe, and the view survived. It is a myth that medieval people believed that their world was flat. On medieval maps, the world is drawn as a circle. For the Greeks, the earth had four parts, of which only one was believed to be habitable. Within this single part were Europe, Asia, and Africa, the three known continents. This circular shape to the maps is behind one half of their modern designation as T-O maps. The "O" is the overall shape of the world; the "T" component reflects the division of the continents provided by the meeting of the important ancient rivers. The scholar Jean Verdon describes T-O maps as follows: "The disk was divided into two axes in the form of the T, the horizontal indication the Mediterranean and the vertical the Nile and Tanaïs. The T marked off three parts: the upper half was Asia, the lower left quarter Europe, and the lower right Africa" (Verdon 2003, 133–134). On the maps, which placed north on the left side, Asia was often twice the size of either Europe or Africa. There was also a theological component to the T-O map. The "T" symbolized the cross and crucifixion of Christ and the ultimate redemption of the world. These three continents were also those that were divided among the three sons of Noah. The medieval world was a biblical one, and this is expressed in its maps.

Other types of maps included "Zonal Maps," which depicted the five zones of the earth—two frigid (near the poles), two temperate, and the middle zone near the equator—as horizontal bands on the circular earth. Only the temperate zones were thought to be habitable. It was not uncommon for the earthly paradise to be placed in the eastern corner near the top of the world. Most of the information available to medieval cartographers was literary. As a result, maps were mostly an impression of a verbal description. These maps had no topography and depicted only sparse physical features. Only place names were inscribed on the map surface.

Cartographers also employed accounts from travelers, and this was likely the only everyday experience of most people with cartography. The return of crusaders brought a wealth of new information, and, as Europeans explored areas in the east, new sites were placed on T-O maps. Around 1100, near the time of the First Crusade, the city of Jerusalem formed the center of the world on new maps. Placing this important city at the center was a reflection of its importance in medieval thinking because maps represented not the precise location of cities but rather their allegorical importance. Thus, many medieval maps were used exclusively by scholars and were lessons in theology and hierarchical order. The most important locations on earth sat at the center of maps, and locations of secondary importance were placed further away from the center.

Like Christian T-O maps, early Islamic maps are general representations of the entire world, but with less allegory. T-O maps often placed Jerusalem at the center, and, similarly, Islamic maps took the area of the modern Middle East as their focus. In the late ninth and early tenth centuries, Abu Zayd Ahmad ibn Sal al-Balkhi (d. 934) established a school of geography where atlases depicting the world, the Mediterranean Sea, and the Indian Ocean and maps of more than 17 Islamic provinces were produced (Short 2003, 70, 72).

In 1410, the geographical writings of Claudius Ptolemy known as *Geographia* were recovered for the first time since antiquity and translated into Latin. Ptolemy, who is better known as an astronomer than a geographer, wrote his works in the second century c.e. He posited a spherical earth, which he then replicated on the flat map. The earth was divided into 360 degrees of latitude and longitude. Mathematical methods used to achieve this schematic were unknown in the medieval world and would be developed from the study of Ptolemy's work. The size of his map was approximately 20 percent too small, and he joined the tip of Africa with Asia. For the area encompassing Europe, the map looks remarkably modern, but for other regions the map is speculative. Despite the proliferation of maps, people did not use them as an aid while traveling, which "was done with assistance of human guides" who had learned their trade through apprenticeships and practical experience (Edson 1999, 4–6, 10).

Around 1300, the first sea charts begin to appear. These charts, known as *Portolano* charts (from the Italian word for "easily available"), were inscribed not with lines of latitude and longitude but rather with a web of compass and wind directions. Wind direction was divided into 8 primary winds, 8 half-winds, and 16 quarter-winds. The known directions to which a compass might point were replicated on an existing map of the ocean and shorelines. Navigators simply had to align their compass so that it matched the desired direction. Standard orientation of maps was a result of *Portolano* charts. North was always placed at the top of the map because it was easier to set the compass that way and then alter course accordingly. Soon after the introduction of the *Portolano* charts, schools of naval cartography were established, and the production of sea charts expanded greatly between 1300 and 1600. These charts were very valuable pieces of navigation equipment and were guarded as state secrets. When an enemy ship was captured, one of the first things the captors plundered was its collection of charts. By the middle of the sixteenth century, *Portolano* charts were collected in the first *Portolano* atlas. The books were coveted not only by sailors but also by wealthy collectors, for whom special fancy versions were created (Martin 2005, 46; Short 2003, 62–64).

SHIP BUILDING

Our knowledge of the history of shipbuilding is complicated by a lack of written sources and sketchy archaeological remains. It is made doubly difficult by the use of technical language that refers to different techniques for laying planks and different styles of ships. In what follows, much of this specific language is ignored for the sake of expediency.

The medieval period began with uncovered ships with a single mast carrying one square sail, and it ended with covered multiple-deck ships with two or three masts hoisting a variety of sails. According to one assessment, "ship-building rivaled architecture as the most advanced technology of the Middle Ages" (Hutchinson 1994, 4).

Viking ships were built using a clinker technique, meaning that the hull was fashioned out of overlapping planks that ran the length of the ship. Sometimes a frame was added for additional strength, but not always. This process worked well on smaller ships, those less than 100 feet in length, because past this length it became too difficult to join the planks. Where the planks overlapped one another, they were planed down to almost the width of a piece of paper so than the outer hull remained smooth. Seams were made watertight with a mixture of animal hair (most often horse or cow) soaked in tar. Boards were joined with rounded iron nails hammered from the outside of the ship. The technique proved very popular. Around the year 1066, English ships, for example, were almost all clinker-built. By the thirteenth century, these Viking-style ships (long and narrow, with a single square sail) were common in northern European waters.

Over time, two structures called castles were added to the stem and stern (front and back) of the ships. A castle is a raised platform from which sailors could observe their surroundings or attack other ships. Later, castles held cannon and gunners. On cargo ships and exploration vessels, the castle held crew quarters. At first the castles were open-air structures that looked remarkably like the top of towers found on traditional stone castles. In the middle of the fourteenth century, ship design changed, and the hull became higher and wider. There was now a distinct stem, or bow, of the ship, and steering was accomplished through a sternpost rudder. Ships still carried a single sail, which was sufficient for northern European travel.

A different variety of early medieval ships was the cog, characterized by a relatively flat bottom and a specific stem and stern. Initial references to the cog date from 948 C.E.; it quickly became the principle ship of the Hanseatic League of northern European traders. A cog had high sides and a relatively straight stem and stern; the mast was positioned slightly forward of center, and there were rectangular castles both stem and stern. Building a cog was somewhat more complicated than constructing earlier types medieval ships. The keel (the main support that ran under the bottom of the ship for its entire length) of a cog was usually made up of three joined parts: a centerpiece and the two ends. The stem and stern pieces of the keel each had two parts, an inner and an outer, which were joined only after the planking had been fixed to the inner. Wooden planks of a cog were flush mounted (had no overlap), and this required a great deal of skill on the part of the joiner. To achieve the smooth sides, augers were used to drill pilot holes for the nails. Mallets hammered the nails into position, and tools to bend the nail flush against the hull were also used (Hutchinson 1994, 15–18).

In the Mediterranean areas of Europe, shipbuilding and design were different from either the clinker-built or cog ships of the north. In contrast, southern ships were caravel built. A caravel ship is defined by having its flush sides joined edge to edge lengthwise along the keel over a preformed frame. The seams were sealed with a waterproofing caulking compound. Unlike their northern counterparts, Mediterranean shipbuilders could produce smooth planks with square edges because they possessed one piece of technology absent in the north in the early medieval period—the saw. Mediterranean ships used a lateen sail (triangular) on a single mast.

Caravel ships required a strong frame on which to attach the planks, whereas the northern clinker ships formed the hull out of planks supported by cross-members added at a later stage of construction. Northern ships had a distinct advantage over southern models: the clinker construction made them more watertight. Southern ships, however, could be built to any desired length. Caravel ships required a massive lateen sail, which greatly reduced the space available at the stem for a castle; thus, only stern castles could be built on early examples of caravels. This lack of crew space made these vessels unsuitable for lengthy journeys. By 1200, however, these ships had two masts and castles on both stem and stern.

When cog ships from the north began venturing into the Mediterranean, a change in design occurred. By the middle of the fourteenth century, the two designs had merged: a caravel construction hull with one mast carrying the original lateen sail mated with a second mast holding the square sail of northern designs. The new hybrid ship had castles both stem and stern. These ships were known as carracks and were the forerunners of the famed galleons of the sixteenth and seventeenth centuries. The most famous example of a carrack is the *Santa Maria* sailed in 1492 by Christopher Columbus. Until the fourteenth century, the region of origin of a ship could be determined by the ship's shape and sails. After this date, however, the merging of design elements made almost all ocean-going vessels practically indistinguishable, with the carrack becoming the ship of choice. Estimating the size of any medieval ship in recognizable modern terms is made difficult because medieval people did not measure their ships in terms of water displacement. Rather, ships were categorized by the number of wine barrels (tuns) they could hold. For example, cogs ranged in size from 500 to 1000 tuns, and Viking ships are known to have been much smaller, with a capacity of 100 tuns.

THE COMPASS

Invented in China and widely used in Europe by the thirteenth century, the compass greatly enhanced exploration at sea by allowing navigation even on cloudy days because the needle of the compass always points in a northerly direction. The thirteenth-century Italian traveler Marco Polo is purported to have returned from his Chinese adventure with a compass.

Medieval depiction of ships. The Pierpont Morgan
Library, New York. MS G. 55 f. 140v.

Prior to the arrival of the compass, European sailors navigated by
a variety of techniques. One method was known as "dead reckoning." This involved estimating a ship's speed and course to determine
distance traveled, which was then traced on a chart and the ship's
direction altered if needed. "Educated guess" best describes this
method. Another navigation approach relied on the recovery of ancient
knowledge and focused on observing stellar positions. The Roman
astronomer Claudius Ptolemy's *Geography*, sometimes called *Geo-
graphick Syntaxis*, contained 27 maps and, after being translated in the
early fifteenth century, quickly established itself as the geographical
reference book of the medieval period. Included in the *Geography* were
listings for more than 1,000 stars and the mathematics needed to tri-
angulate their position with respect to the observer. Thus, ships could
navigate by noting their location as judged by the stars. The North

Star often served as a fixed reference marker. On cloudy or foggy days, however, this procedure was useless. As a result, early medieval ships rarely ventured very far from the visible coastline (Blake 2004, 15–16).

Medieval compasses were made from loadstone (magnetite, a naturally magnetic ore). The earliest compasses consisted of nothing more than a magnetized needle stuck through a piece of straw floating in a bowl of water. Vikings may have been early users of a floating magnetic needle to determine direction. The earliest recorded European use of a compass dates from 1180 and is found in Alexander Neckham's *De Utensilibus* (On Instruments). Later modifications to the compass included attaching the needle to the bottom of the bowl so that it only rotated in circles. Next, a card adorned with the four directions of north, south, east, and west (later 32 other divisions were added) was attached to the base to make readings easier. Around 1300, the first modern-looking compass was produced. It was only when mounted on a directional card and then combined with charts that the compass became anything like the navigational aid known today. Although it was known that the needle pointed not at true north but rather at magnetic north, which wanders slightly, the compass was an indispensable navigation tool. Only in 1600, thanks to the research of the Englishman William Gilbert, did Europe learn that the Earth itself was a giant magnet.

How the compass (and magnets) worked was a debated topic that extends back to classical antiquity. Magnets were known as a curiosity in the ancient world. Greek philosophers, and later their Roman counterparts, attempted to explain the attraction and repulsion of magnets. There were two main interpretative theories: the first posited that an unseen emanation exited the magnet and pulled or pushed metal, the second that a kind of sympathy existed between the magnet and the piece of metal that linked them in some unknown way. Members of the first school of thought included the famed philosopher Plato (428–347 B.C.E.). Atomists such as Democritus (ca. 460–370 B.C.E.) and Epicurus (341–270 B.C.E.) explained magnets' functioning by positing an interaction among tiny particles or atoms. Opposed to physical explanations were Pliny the Elder (ca. 23–79 C.E.) and the physician Galen (129–199 C.E.), who believed that a kind of soul was present in the magnet and that it drew metal to the magnet by immaterial means. This was also the view of Aristotle. After the fall of the Roman Empire, the writings of Pliny, Galen, and Aristotle survived and set the tone for learned discussions of the magnet in the medieval period. Medieval philosophers interpreted the action of the magnet as an aspect of God's divine providence (Jonkers 2003, 39–41).

THE RUDDER

The earliest medieval ships in the north steered by using a single side oar most often attached on the starboard side of the ship (right side of a forward-facing ship). Vikings, for example, steered their ships with a single side oar. In the Mediterranean, slightly different methods were employed. Southern European ships had two steering oars on the port (left side of a forward-facing ship) and starboard sides. For these long and low ships, steering oars worked very well. The side rudder was also sufficient to steer cog-style ships until the late twelfth century (Unger 1980, 130, 141).

As ships grew in size, the need for a different method of steering became evident. Since side rudders had to be fashioned from a single tree trunk, there was a limit to their length. The size of newer ships simply exceeded the capacity of the side rudder. A sternpost rudder became a necessity because the alternatives were insufficient to steer a ship of great bulk. Without the ability to steer these larger ships (round and broad), their cargo capacity would be rendered irrelevant. The key development in ship steering was the sternpost rudder. The sternpost is part of the keel that extends out of the water at the stern (rear) of the boat. The earliest possible depiction of a sternpost rudder dates to around 1180. However, this relief, found on the front of an English church, is not definitive, and most scholars place the arrival of the sternpost rudder in the thirteenth century. By this time, the stern rudder was well on its way to replacing the alternatives and quickly became the norm on all northern European ships. Sternpost rudders were introduced on Mediterranean vessels in the fourteenth century.

The historian Gillian Hutchinson argues that side rudders have been unfairly criticized in favor of the sternpost rudder. The side rudder, she suggests, "is balanced and exerts pressure on either side of its center of effort. It is therefore more efficient, causing less drag and needing less effort to keep it at the required angle, than the sternpost rudder which is unbalanced because it swings from its leading edge" (Hutchinson 1994, 50). Despite Hutchinson's eloquent defense of this earlier technology, most historians hail the sternpost rudder as a major innovation and one that forever changed medieval sailing.

SAILS

Vikings used sails made from woolen cloth and secured with rope fashioned from walrus hide. In England, wool was often the sail material of choice until it was superseded by sailcloth made from canvas. The most basic sail shape is the rectangular or square sail. The square sail is hung from the crossbeam (known as a yard) on the front side of the mast. The sail catches the wind on one side only. According to the historian Alan McGowan, "This type of sail was developed in the northern seas where winds of gale force are common and frequently last for days on end" (McGowan 1981, 9). In this environment, the goal was to employ as much wind power as possible. As an example, early Viking ships used a sail shaped like a rectangle. The mast on these ships was very tall—often between 11 and 14 yards. Propulsion was not achieved by wind power alone because the ship carried many rowers (oarsmen). The performance of square sails is greater than historians had previously believed. Recent scholarship suggests that the wind did not necessarily have to be blowing directly into the sail for the ship to be propelled forward. Captains of vessels could alter the angle of the sail slightly so that it could take to the wind from several shallow angles.

The calmer, more variable winds present in the Mediterranean required a differently shaped sail known as the lateen. The advantage of the lateen was that it permitted the ship to travel much closer to the wind (as near as possible in the exact direction of the wind). Arab sailors invented the lateen in the second century. Ships could also leave port whenever they wanted and did not have to wait for a favorable wind. Basically, the lateen is a triangular sail. The sail is attached to the yard on one side and to the mast on the other. The yard pivots freely to take the wind from any direction and power the ship forward. (It helps to think of the last sail on tall ships.) As a result, ships were much more maneuverable and could be sailed in any wind conditions and at any time of the year. Ships could also carry much more cargo because the space that had been often used to house rowers could now hold materials. The new empty cargo hold could also carry cannon. This turned ships into battleships, a more powerful weapon for war. Some downsides of the lateen included the extreme difficulty of moving the sail from one side of the ship to other even in calm days, let alone in a storm. Also, strong winds from directly behind the ship could cause the ship to capsize because of the low level of the sail sat in relation to the ship (McGowan 1981, 9–11).

When paired with the square sail, the lateen sail forever changed medieval maritime travel. The lateen also helped spur exploration because ships could now travel in any wind condition. The lateen allowed ships to depart harbor at a time of their choosing, and the square sail was more efficient with the stronger winds of the Atlantic. The date of the initial combination of the square and lateen sails remains unknown, although many historians suspect that merchants from Genoa created the first multimast and multisail ships in the late fourteenth century.

RUMORS OF FOREIGN LANDS AND PEOPLE: MARCO POLO AND KING PRESTER JOHN

Rumors and mythologies associated with foreign lands were often the impetus for medieval exploration. Even when the existence of formerly mysterious lands was no longer questioned, accounts of these lands still served to spur on adventurers. The earliest mythological land that explorers sought was Atlantis. First mentioned in the dialogues of the Greek philosopher Plato, the lost continent of Atlantis was believed to reside somewhere in waters to the west. Unsuccessful attempts at locating it were launched in the Mediterranean and in the Atlantic. A rumor that turned out to have a grain of truth concerned the unknown land mass at the bottom of the world known as *Terra Australis Incognito.* Again a Greek writer, Aristotle in this case, is the source of the legend. For reasons of global symmetry, Greek philosophers posited the existence of a large continent in southern waters to balance the lands known to exist in the north (i.e., Europe, Asia, and Africa). The search for *Terra Australis Incognito* continued well into the medieval period and was proven false only in the eighteenth century. Other southern land masses were discovered, such as Antarctica and Australia, but these were not where the mysterious continent was though to be located (Martin 2005, 78, 102).

Mysterious lands must be populated with mysterious people. It was commonly believed that the different climate and geography of non-European territories had led to the development of different races of people. Originating with ancient authors' accounts of unknown exotic places, most often "the edge of the world," fables of unusual creatures and inhabitants survived and flourished in the medieval era. Such descriptions were invoked to emphasize the possible dangers posed by foreign travel and by journeying too far from Christian civilization. For example, cannibals were thought to live in many unexplored lands. Descriptions of monstrous humans included claims that they possessed exaggerated or missing physical features such as enormous ears, a single foot, a face on the chest, and giant stature. A famous example of legendary giants are the Amazons, a race of large women believed to live at the end of the known world. The myth dates to antiquity but was given new interpretation in the medieval era, when Amazons were depicted as living in quasi-convents. Whereas the Roman naturalist Pliny the Elder described races of giants in his writings, medieval scholars agreed that most giants were extinct (they had perished in the Flood) and that those who survived were widely dispersed in the world.

Unseen races could also have biblical and theological origin, such as Gog and Magog, a race that was thought to exist somewhere near the Caucasus Mountain range (between the Black and Caspian Seas and comprising such modern countries as Georgia, Armenia, Azerbaijan, and Russia). Medieval thinkers took Gog and Magog to be one of the Ten Lost Tribes of Israel, whose arrival in Christian Europe would signal the End of Days. Almost all of these mutant humans were in this deplorable condition because of some moral lapse for which God had exacted punishment in the form of perpetual difference from the rest of the pious Europeans. This is seen most readily in medieval descriptions of "Wild People" who existed without morals at the margins of civilized society. They were believed to live in trees, caves, and forests. Wild People hunted with clubs or uprooted trees; they rode stags rather than horses and were recognizable by their long, shaggy hair (Campbell 1988; Friedman 1981).

Long before the arrival of the Venice-born merchant Marco Polo (ca. 1254–1324), China had a trading relationship with the Roman Empire.

This was conducted along the famed Silk Road. After the Roman Empire fell, Chinese contact with markets in the west dropped significantly and rebounded only under the Mongols in the thirteenth and fourteenth centuries. Even with trading re-established, few European merchants ever traveled the entire distance to China. Rather, intermediaries, such as Arab sailors, were used to maintain a relationship between Asia and Europe.

In 1271, at the age of 17, accompanied by his father and his uncle, the young Marco Polo began his historic journey into the mysteries of the East. The trip lasted 24 years and covered much of Asia, including parts of China controlled by the grandson of Genghis Khan. The Polos reached Shang-tu, the seasonal home of Kublai Khan, leader of the Mongols, in 1275. Kublai took an immediate liking to the Polos, and for the next 17 years they served as advisers and administrators to the Khan. As part of his duties for the Kahn, Polo visited much of China and parts of India. In 1292, Polo, his father and his uncle began the long journey back to Venice, finally reaching their destination in 1295.

Once back in Venice, Polo set his experiences on paper. The result was *Description of the World* (1298), one of the earliest accounts of Chinese culture written in Europe. At first, the fantastic tales of Chinese wonder were seen as fiction and exaggeration. Only after his death was Marco's book taken as factual reporting. The account of China provoked interest in far-off places, and the geographical information Marco provided within the pages of his book was the most important source for those making maps of Asia for generations. The book also sparked European interest in unknown lands and the riches held in them.

In addition to the financial motivation for exploration, there was a religious incentive, too. Throughout the medieval era, rumors persisted about a Christian king somewhere in the East (the exact location was the subject of much speculation) who was named Prester John. It was believed that John was descended from one of the Magi present at the time Christ's birth. The long search for John and his kingdom has been described as "one of the greatest romantic enterprises of the Middle Ages." Early versions of the legend may be traced to 1145, when the last Christian holding in the Holy Land taken during the Crusades fell back into Muslim hands. Desperate for relief, Bishop Hugh of Jabala traveled to gain an audience with Pope Eugenius II with the intent of securing new troops. It was during one of their meetings that the Bishop told Eugenius about Prester John. This king was believed to be a Nestorian Christian (referring to a heresy, outlawed in 431 c.e., that preached that Jesus had two bodies, one divine and the other flesh). Medieval gossip had placed the Nestorians in an unknown eastern location. Prester John and his Christian army were now coming back to western Europe and were eager to join the Crusades and secure Jerusalem. Unfortunately, obstacles always seemed to prevent John's timely arrival, and no one had actually seen him.

The mysterious king seemingly did not wait to be contacted, and, in 1165, John purportedly sent a letter to the Byzantine Emperor Manuel I Comnenus in which John described the pious virtue and merits of his kingdom. In the letter, John did not provide an exact location of this modern Eden over which he ruled, but he did state that it lay among the Three Indias (parts of modern India and the east coast of Africa). Although it would later be proven a clever forgery, during the twelfth century the letter was accepted as fact, copied, and translated, and hundreds of versions were circulated throughout Europe.

In 1222, following the Fifth Crusade, reports circulated that a certain King David (either the son or the grandson of Prester John) was on his way from an undisclosed eastern location to help the besieged crusaders. Marco Polo contributed to the legend in his account of China, in which he claimed to have personally met Prester John. Perhaps on the basis of this report, Prester John was, for a brief time, identified as a member of Genghis Khan's family. It was then reported that Prester John was living near Ethiopia. Prince Henry the Navigator (1394–1460), a famed Portuguese explorer, had heard the legend of Prester John as a small child, and the tale never left his mind. As a young sailor, Henry was determined to find Prester John or, if not him, his kingdom and his descendants. Henry sent ships of exploration to the African coast with the double duty of seeking trade and Prester John. King John II of Portugal (ca. 1450–1500) also sought Prester John. John II instructed his explorers that, as they scouted trade routes to India, they were also to search Ethiopia for Prester John. The fruitless search would continue for another century before finally being abandoned (Beckingham and Hamilton 1996).

6

TECHNOLOGY AND WARFARE

Images of chivalrous mounted knights in radiant armor galloping across fields of grass to engage the enemy in a series of jousting attacks often characterize our idea of medieval warfare. While these images are romantic and the stuff of movies, the reality of these battles was much more gruesome. Combat was brutal because it tended to maim rather than kill outright. Soldiers frequently died of their wounds after suffering terribly. In this kind of close fighting with a sword, a person was only a piece of meat. Any chivalry and social exclusivity that had existed in medieval battles did not survive the invention of gunpowder and gunpowder weapons such as cannon and firearms.

This chapter examines the technology and science behind medieval warfare. First, the laborious effort required to make armor is revealed. A knight without a sword was no good to anyone, so swords are considered next. While medieval warfare often relied on brute strength, there was a science to it. Thus, military mathematics and kinematics (the science of motion) are discussed in this chapter. Many of the confrontations in this era occurred as lengthy sieges involving machines constructed exclusively for that purpose; these are analyzed in their own section. The arrival of gunpowder in thirteenth-century Europe forever changed the nature of battles, weapons, and defensive architecture. Cannon, firearms, and gunpowder itself are all addressed in separate sections. Last, the measures taken by towns, cities, and castles to ensure that they survived the medieval period are explored.

ARMOR

Armor defines medieval knights and contemporary depictions of battle. The practice of wearing armor comes from antiquity, where Greek and Roman infantry wore armor consisting of a helmet and a breastplate made from bronze. After the fall of the Roman Empire, western Europe was besieged by warring Germanic tribes, who had defeated the Romans without wearing armor, and, as a result, the practice of wearing protective garments faded. In contrast, the eastern half of the Roman Empire, known as the Byzantine Empire, continued to use metal armor in its confrontations with invaders. Vikings, too, eschewed armor and were feared even more because of their choice to fight without bodily protection. However, the majority of Europeans who engaged in battle from about 600 c.e. on did wear some armor. Indeed, owning armor became a prerequisite for entry into the ranks of knights. In the seventh century, the Visigoth king Erving made it law that most warriors should wear armor and that all should carry shields. Two hundred years later, Charlemagne (ca. 742–814) ordered that nobles and owners of a certain amount of land must have their own armor, specifically a byrnie (a long shirt of chain mail) (DeVries 1992, 56, 59; Nicholson 2004, 106).

No instructional manual for making armor has survived from the medieval period. Historians have relied on secondary accounts and contemporary illustrations to piece together the techniques and processes involved in manufacturing both mail and plate armor. Similarly, few tools of the armorer survive, but, from those that are extant, it is possible to reconstruct a likely scenario of the armor-making process. Not until around the fourteenth century did the design of armor worn by knights begin to change. Crusading armies of the twelfth and thirteenth centuries went to war outfitted in much the same way as their counterparts in the sixth century, wearing shirts of mail and rounded, open-face metal helmets. Mail was the dominant armor for centuries. Its relative comfort and effective protection satisfied the defensive needs of European warriors until the development of such weapons as the crossbow necessitated more sophisticated alternatives. It was not until the late thirteenth century that plate armor began to appear—and even then it appeared in the form of attachments to existing mail armor rather than as an entire suit.

Upon its arrival, plate armor quickly superseded mail, at least for those who could afford it. Only in the thirteenth century were masks, solid pieces of metal with vents and eyeholes, added to existing rounded iron helmets. By 1220, the helmet, worn over a mail hood, included ear and neck protection. The completed helmet was known as the "Great Helm." The enormous cost of the Great Helm restricted its use to mounted knights of aristocratic standing. Infantry wore the older style of rounded helmet without a facemask because a mask would have been too hot and cumbersome in hand-to-hand combat. Armor for the exceedingly wealthy, such as kings, could be adorned with gold and jewels. Other personalized touches included elaborate engravings. Often this decoration was then blackened in the background to make the design stand out even more. The prohibitive cost of armor and helms meant that lesser nobles and less wealthy warriors continued to rely entirely on the protection offered by mail for centuries after the introduction of plate armor.

The sheer quantity of mail produced (it was still being manufactured into the late seventeenth century) indicates that its construction was relatively simple and had become very refined. Only the final component of linking the fabricated rings into the finished mail shirt fell to the master craftsman, while all other jobs were done by less skilled persons. Hence, mail manufacturing was an early example of an assembly line. There were at least two potential methods for producing the iron rings that would be linked to produce mail. Closed rings were likely stamped out of iron sheets. A punch was then used to form a more precise ring. Alternatively, open rings were fashioned out of iron wire. The production of wire was labor intensive. In one technique, an iron rod was drawn through smaller and smaller holes until the desired thickness was achieved. Next, the iron wire was likely coiled around a dowel and then cut into individual open rings. Each ring (or link) was about half an inch in diameter, and there were often 30,000 rings in each mail shirt. Another process involved cutting thin strips from iron plates and working them until they met the required measurements. A combination of the two methods took cut iron slivers and pulled them through progressively smaller holes. Much of this process could be done cold, but every so often the armorer had to soften the metal by heating it until red hot and then allowing it to cool somewhat before proceeding with the manufacturing. To ensure maximum protection, each ring was linked through four others before being closed with a rivet. Sometimes an armorer would alternate open and closed rows of mail. For particularly strong armor, two rings would be used in place of one. From the twelfth century on, advances in metalworking carried over to advances in armor. Chain mail now commonly consisted of two or three layers (Edge and Paddock 1988, 176; Pfaffenbichler 1992, 56–57).

Plate armor appeared first as small pieces worn over mail to protect the shoulders and knees. Complete plate armor suits were in regular use by the early fifteenth century. To produce a full set of plate armor required the collective talents of several craftsmen. First, the armorer would fashion the plates; next, the polisher would remove any residue and provide a brilliant sheen to the plates; and last, the finisher would assemble the suit by attaching straps and cushioning. Raw plates of armor were hammered out of pieces of iron and, later, steel. At first this labor was done entirely by hand. The main tools were the hammer, anvil, vise, and chisel. There was also a tool called a *nailetoules* that was used to close the rivets holding the various pieces of metal together in a finished section of armor. With the diffusion of waterwheels, however, the use of hammers operated by waterpower became the preferred technique for accomplishing this task. Once a rough shape had been created in the now-thin metal sheet, the real craft began with the curving and molding of the flat piece of metal into a wearable protective garment. While much of the work was done with cold metal, some finishing touches like edges required that the metal be heated until red-hot. A truly skilled armorer could vary the thickness of the plates to provide protection where it was needed most. For example, the center of the breastplate was often thicker than the surrounding areas because early armor was designed so that it offered protection from a glancing blow. Indeed, the armor was built with rounded edges and raised surfaces that produced a glancing blow if struck by a sword in battle.

Iron was the chief material for making armor. To produce very hard armor, however, required the production of carbon steel. In this process, raw iron was wrapped in animal skin and then heated for a considerable length of time. As the piece heated, carbon soaked into surface of the iron, turning it into steel. The longer the heating lasted, the greater the amount of steel produced. Another, quicker, method consisted of placing iron surrounded by charcoal into a furnace for a time; this deoxidized some of the iron, making it into a natural steel. This thin layer of steel could be hammered off and forged into blanks of steel, or the steel layer could simply be used to strengthen the inner iron core. The hardness of the steel could be increased by rapid cooling, such as by submerging the red-hot plate under ice water. While this did result in hardening the steel, it also tended to make the steel very brittle—not a very desirable trait in armor. To counteract this fact, the piece was slowly reheated after being cooled. This process is known as tempering. Tempered steel, however, was available only in the later medieval period (Ffoulkes 1988; Nicholson 2004, 108–109; Pfaffenbichler 1992, 62–64).

SWORDS

The sword was the chief weapon of medieval warfare. The historian Jim Bradbury calls it the "medieval weapon *par excellence.*" Similarly, Kelly DeVries states that the sword was "the most celebrated" of all medieval weapons even though the majority of fighting men in early medieval Europe used spears and axes. DeVries explains that there was a mystical, awe-inspiring aspect to the sword: leaders and true warriors carried swords. Like other pieces of medieval military technology, swords trace their origins to antiquity. Roman infantry adopted swords from the Greeks and the Etruscans. The Romans used a *gladius,* a short sword meant for stabbing. Medieval swords were expensive, a symbol of status and power for their owners, and were usually passed down from one generation to the next.

Swords of the early medieval period were forged out of iron. Repeated hammering of heated iron into the rough shape of a sword began the lengthy process. Once the raw sword was completed, it would be ground using progressively finer stones until the edge was razor sharp. After the blade was completed, the handle and guard were fixed in place. The process was similar all over Europe, with the quality of the original iron ore being the only major difference in swords produced in various countries. Bog ore (deposits of iron that accumulated in swamps or bogs through the oxidation of iron carried in solution) was a common source of iron; the use of bog ore dates from the fourth to the first centuries B.C.E. Bog-ore iron swords were longer and tended to be thinner and more flexible than the bronze swords of antiquity. A sharp double edge characterized the early medieval sword, which was a weapon of slashing.

The best swords were constructed using a process known as pattern-welding. Craftsmen using this technique, invented by the Celts, could take nearly a month to make a top-quality weapon. A pattern-welded sword blade was a composite of many parts. A craftsman first began with many thin rods of iron with their surfaces covered in a layer of steel. Next, the rods were heated and twisted together to form the core of the sword. The edge of the sword was a billet of pure steel shaped into a V. Forging the white-hot core and edges together produced the finished sword. The softer iron core of the sword prevented it from being shattered in battle. Pattern-welded blades continued to be made until around the eleventh century. After this time, a technique appeared for manufacturing homogenous steel, in which craftsmen introduced carbon to heated iron bars and hammered off the layer of steel; they then combined these thin sheets into a thick block of pure steel, which was then forged into a sword.

The overall design of swords did not change much during the period between 800 and 1300. The slashing motion that had been used up until that time and that had been so effective against chain mail became highly ineffective against a knight enclosed in the new plate armor that appeared in the fourteenth century. Swords changed to meet this new innovation. The killing method of choice became a thrusting action that would penetrate the armor. Swords were therefore made shorter and with a strong sharp point. In other words, there was a return to Roman-style swords. With the rise of gunpowder and handheld weapons, the usefulness of the sword for attack and defense was greatly diminished, and it faded from European battlefields, though it was still worn by gentlemen to denote their social standing (Bradbury 2004, 248–249; Edge and Paddock 1988, 25–27; Nicholson 2004, 103–104; Santosuosso 2004, 134–135). Swords are still used in modern warfare; we know them as bayonets.

SIEGE MACHINES

Attackers attempting to capture a city, town, or castle often faced imposing defensive walls that stood between them and their prize. The most effective way to besiege a castle or fortified town was simply to surround it, cutting off all supply lines and starving the inhabitants into submission (Santosuosso 2004, 167). However, invaders in a hurry relied on various machines to achieve their goal. Less exotic siege machines included ladders and wooden towers that enabled the attackers to get over the walls of the enemy fortification. A belfry, for example, was a wheeled tower assembled on site to a height matching that of the defensive wall. The belfry was pushed or pulled close to the wall so that the attackers could engage the enemy on the same level. Rather than going above the battle, some techniques required one to go beneath it. Mining involved building a tunnel under a defensive wall and replacing the stones with wood and then firing on the wood to topple the structure.

When these relatively simple technologies did not work, more awesome machines were used. These siege machines hurled a variety of projectiles toward defensive structures in order to bring them down or to punch large enough holes into the walls so that an invading force might enter a formerly secured city or castle. The pre-gunpowder terror weapon of choice in the medieval period was the trebuchet. This device likely originated in China between the fifth and third centuries B.C.E. and came to the Europeans via the Arabs. A trebuchet was a long beam that rotated and was attached to a supportive base by means of a scaffold tower. How the beam rotated was the key to the trebuchet's function. The pivot point for the beam was off-center in a ratio of about 6:1 for lighter models and 3:1 for the heavier variety. For example, a trebuchet of 100 feet would pivot at about the 84- or 75-foot marks, respectively, depending on its weight. Attached to the longer end of the beam was a sling or basket into which was placed the projectile, most often a large stone. On the shorter end of the beam were placed about 100 ropes, which, when pulled by soldiers, activated the trebuchet and caused the projectile to be hurled a distance between 100 and 150 yards. A trebuchet was a valuable piece of military hardware that could be dismantled and taken with an army after it left the field of battle.

When this weapon was introduced in Europe is a matter of historical debate. Some scholars believe it was in use around the sixth century, while others suggest that it was unknown until crusading armies saw a trebuchet in the Holy Land in the late eleventh century. Historians such as Kelley DeVries suggest that crusaders used a trebuchet in 1147 to take Lisbon from the Muslims. On this occasion the trebuchet reportedly fired at a rate of 250 stones an hour. However, the historian Helen Nicholson argues that because the description of the weapon used is vague, more evidence is required before it can be called a trebuchet with any certainty.

Similarly, Antonio Santosuosso claims that trebuchets were very uncommon in Europe as late as 1300 and that scholars should be cautious when commenting on their proliferation. Nonetheless, there are confirmed uses of trebuchets in medieval Europe after 1147.

The accuracy of trebuchets was only as good as the skill of the men who activated them by pulling on the ropes with sufficient force to get the pivoting arm moving. This unpredictability initiated a search for a more dependable source of activation. Thus, the counterweight trebuchet was born. Like that of the trebuchet itself, the date of the first use of a counterweight trebuchet is open to scholarly debate. Some historians suggest the siege of Zevgminon in 1165 for its first appearance, while others suggest an earlier although unsubstantiated use by eastern armies in 1097 at the siege of Nicaea. This version of the weapon was, in all respects, identical to the older incarnation except that the force of motion generated by the men pulling was replaced by a counterweight. Counterweights were massive, weighing somewhere in the neighborhood of 9,000 to 30,000 pounds. The extra weight created extra momentum and force of motion that allowed these new trebuchets to hurl heavier projectiles greater distances, sometimes as far as 330 yards. Trebuchets were very effective at smashing defensive walls. They were also the forerunners of modern strategic bombing in that a trebuchet could be used to destroy the morale of defenders by hurling not only rounded rocks but also fire and more gruesome things like diseased animals, severed body parts, and, occasionally, live captives at opponents (DeVries 1992, 133–134; Nicholson 2004, 92,

Depiction of a trebuchet in warfare. The Pierpont Morgan Library, New York. MS M. 638 f. 23v.

95–96; Santosuosso 2004, 168). By the middle of the fourteenth century, it was clear that the sun was setting on the era of the trebuchet. Advances in gunpowder weaponry quickly made the former terror weapon increasingly obsolete, although trebuchets would prove a lasting companion to the new cannon, and the two used together produced a powerful one-two punch until cannon secured their place on the battlefields of Europe.

GUNPOWDER

Early authors marveled at gunpowder's explosiveness but did not suggest its usefulness for new weapons. The basic recipe for gunpowder is saltpeter, sulfur, and charcoal in specific proportions. An ideal mixture of ingredients contains 75 percent saltpeter (potassium nitrate), 12 percent sulfur, and 13 percent charcoal. This perfection, however, was notoriously difficult to achieve. Getting the mixture of gunpowder's constituent parts correct was critical for the end result, which was, after all, firing an iron ball or stone with enough force to do damage to the target. Precisely speaking, gunpowder does not explode, it burns, and the rate of burn is highly dependent upon the percentages of the ingredients in the gunpowder. Batches of gunpowder varied in strength until a consistent technique for manufacturing the grain size of the powder was established. This process, known as corning, or granulating the powder (think sugar here), was well established by the fifteenth century in Europe.

Gunpowder may be traced to ninth-century China, where Tao alchemists likely produced it for firecrackers to celebrate the New Year. Later, around 900 C.E., the Chinese developed something called a "fire-lance" that propelled fire and smoke but little else. Bits of shrapnel were later put into the lance. From China, the technology passed into Islamic territories. Muslim soldiers used the noise of gunpowder to startle enemy horses into a stampede. Through contact with areas housing Muslim settlements, medieval Christian Europe acquired knowledge of gunpowder during the thirteenth century (Crosby 2002, 97–99). Initial scholarly assessments of gunpowder's uses were guarded. The Oxford thinker Roger Bacon described a mixture for gunpowder in *Epistola de Secretis Operibus Artis et Naturae et de Nullitate Magiae* (Letter on the Secret Workings of Art and Nature, and on the Vanity of Magic), composed between 1248 and 1267. Bacon was wary of this substance, which seemed to give its user the power of God. Also, around 1267, he warned of the thunderlike properties of gunpowder and its potential to "disturb the hearing." Similarly, the Dominican scholar Albert the Great considered gunpowder in *De Mirabilibus Mundi* (Concerning the Wonders of the World) around 1275. His concern, like Bacon's, was the ability to make "flying fire" and the God-like powers this gave to the user. Noble society, kings and knights, did not embrace gunpowder weapons with open arms. Their place in society was established by their ability to fight in time-honored ways and by their wealth, which allowed them to afford armor, swords, and horses. Gunpowder nullified all that made these people special. A mounted knight was no match for gunpowder-propelled projectiles, which had no respect for feudal status. Eventually, however, even the most steadfast opponents were captivated by the immense power and potential of gunpowder. Further down the ranks, regular soldiers, too, were reluctant to use gunpowder weapons. These weapons that rattled the ground and produced terrifying roars seemed the work of Satan, not God. This aversion soon passed (DeVries 1992, 159–160).

After 1300, gunpowder became more commonly known. (The first recorded instance of gunpowder being used to propel projectiles occurred during the siege of Metz in 1324.) The problem then became one of supply, although, since relatively few weapons used gunpowder, initially the problem was not severe. By the early 1400s, however, the number and caliber of guns rose dramatically, as did the demand for the powder that fired them. So prevalent were gunpowder weapons in the fifteenth century that contemporaries wrote of battlefields smelling of burnt powder and smoke (Nicholson 2004, 97). However, after its first century of use, roughly 1325 to 1425, gunpowder still was more smoke than bang. This new terror weapon did not live up to the tremendous hype that had accompanied its introduction. The trebuchet was more reliable and more trusted. While the use of cannon on the battlefield against opposing armies was a limited success, their usefulness during sieges revealed their enormous untapped potential. By 1450, nearly all sieges relied heavily on gunpowder weapons. As the price of gunpowder began to fall, in the late fourteenth century, the use of cannon proliferated. Cheaper gunpowder meant that more of it could be used for propelling projectiles of considerable size. In the early 1400s, a kind of arms race took place, with several territories each producing increasingly larger cannon that could shoot larger and larger stones or balls. Hundred-pound rocks, though intimidating around 1400, were superseded as fast as larger cannon could be constructed. The largest cannon still extant today holds the record. It was capable of shooting a ball weighing more than 1,500 pounds (Hall 1997, 58–59).

CANNON

Once the potential of gunpowder as a propellant became known in medieval Europe, weapons that used it soon followed. The first references to cannon are from the late thirteenth and very early fourteenth centuries, although many historians question the validity of these reports. However, that cannon were regularly used by the 1330s is accepted by most scholars. The first European picture of a cannon, dating from 1326, shows a pear-shaped vessel. It is depicted firing not a cannonball but, instead, a large arrow (Crosby 2002, 112–113).

Most early cannon were manufactured from forged iron, metal that has been shaped through a cycle of heating and hammering. The easiest method was to solder iron bars together to form a tube, capping one end with a solid iron plate. The composite tube was then strengthened by the addition of a series of iron rings running the length of the barrel and forged into place. The largest known cannon from medieval warfare were the "bombards." They were up to 5.7 yards in length with a caliber as large as 28 inches. Their size made them tremendously difficult to move, and thus these largest cannon were used almost exclusively in lengthy sieges. The grand size of the bombards, which made them such effective weapons, also made them easy targets for the cannon of the defenders. It was soon realized that several smaller cannon could produce the same power as the bombards while being easier to move and much more difficult to destroy. Around 1450, cannon were mounted on wheels, which alleviated the backbreaking task of moving them and absorbed some of the recoil. By the late 1400s, techniques of bronze casting had progressed to the point that cannon of cast bronze became the premier artillery pieces for European armies. The advantage of casting (pouring molten metal into a form) was that the resulting weapon possessed superior strength and was much less likely to explode. The benefit of cast bronze cannon must have been great enough to offset the drastic increase in price (bronze cannon cost three times as much as iron cannon).

Early cannon projectiles were stones and rubble. Later, spherical balls of iron, lead, or stone were used, with more devastating results. Larger cannon like the bombards used stone balls, fashioned by masons, rather than sizable iron balls, which would have been prohibitively expensive. Imperfect casting in the earlier medieval period meant that lead shot was easier to obtain and was the projectile of choice until the fifteenth century. The techniques for casting iron were more difficult (the problem was heating iron to temperatures high enough to melt it), and only relatively simple shapes could be produced by the early 1400s. One such shape was the sphere, and the result was cast iron cannon balls, which became the projectile of choice for battlefield commanders using smaller cannon. They were preferred because such spheres of iron are approximately three times as dense as their stone counterparts, and thus a smaller iron ball could achieve the same impact as a much larger (and more difficult to transport) stone. The availability of smaller, denser spheres meant that cannon could be manufactured with smaller diameter barrels and were more efficient (DeVries 1992, 150–158; Hall 1997, 93–94).

EARLY FIREARMS

The development of and progress in cannon design and manufacturing made the production of handheld models (i.e., medieval firearms) almost inevitable. Initially, the handgun, like early cannon, was a weapon that produced more bang and smoke than actual killing power. Surviving documents dating from the late fourteenth and early fifteenth centuries reveal the great extent of handgun use. The Anatolian Turks had handguns around 1350, and Italians were experimenting with guns in the 1360s. Contemporary reports claim that John the Fearless, Duke of Burgundy, had about 4,000 of these new guns in his army by 1410. Around 1450, all European armies contained at least some firearms as part of their complement of weapons.

Early matchlock arquebuses (or harquebusses) first appeared in battle around the 1450s. The arquebus was both a specific type of firearm and the designation for an entire class of weapons. The name comes from the Middle Low German *hakebusse* (hook-gun), so called because the first versions were so heavy that the gun had to be attached to a carriage by means of a hook. The name stuck. At first, the gun was essentially a metal tube with a wooden stock (Bradbury 2004, 240). The original arquebus had a long barrel of about 40 inches and a quite small diameter bore of approximately 0.6 of an inch. Like a cannon, the gun was loaded through the muzzle and the gunpowder lit by an external device. The first trigger for the arquebus was a slow-match (a piece of smoldering cord) in either an S or a Z shape that was pressed into the powder chamber, causing the gunpowder to burn and propel the bullet.

Arquebuses fired rounded lead bullets through a smooth barrel. (Rifling was a nineteenth-century innovation.) Because of the combination of the smooth bore of the arquebus barrel and the spherical bullet, it was not always easy to predict where the bullet would travel. This contrasts sharply with the accuracy of the modern science of ballistics. Bullets exiting a smooth barrel are much more susceptible to air resistance. What is more, the bullet often travels in the direction of the initial spin it possessed upon leaving the gun. As the historian Bert S. Hall has noted, early firearms often produced a flight path not unlike that of a curve ball thrown by a baseball pitcher or of a sliced golf ball. Modern tests conducted on medieval firearms reveal that their muzzle velocity falls within the range of current handguns and is about half that of modern rifles. However, this velocity quickly diminished due to air resistance. This means that the lethal power of medieval firearms was effective only within a distance of approximately 130 yards. When the unpredictability of the flight path is factored in, it seems reasonable to suggest that of all the bullets fired in medieval battle conditions, fewer than 1 in 500 hit the intended target. Nevertheless, stray bullets found targets, too, and the arquebus found a niche that firearms would never surrender (DeVries 1992, 149; Hall 1997, 95–97, 134–139).

MILITARY MATHEMATICS

Mathematics went hand-in-hand with military operations during the medieval period, even if gunners and commanders did not always realize the importance of mathematics to their attack strategy. Medieval scholars such as Roger Bacon (c. 1219–1292) commented that knowledge of numbers is simple and "is clearly proved by the fact that mathematics is not beyond the intellectual grasp of any one" (Grant 1974, 91, 92). The same mathematics and surveying techniques used to construct the defensive fortresses and castles of medieval Europe were also put to use by those who wished to destroy such structures.

Early versions of cannon tended to be individual items, each having different firing characteristics. In other words, there was no standard pattern to how a specific model of cannon would fire a projectile. Each cannon had to be "zeroed," a process that involved shooting a canon first from a level position and then raising it one degree and firing again. This was repeated until the cannon fired nearly vertically. After each shot, the distance traveled by the projectile was recorded. In 1537, an invention by the Italian Nicola Fontana made the task of determining the exact cannon angle easier. His tool was the gunner's quadrant, which fitted into the barrel of the cannon and held what in modern terms would be a right-angle protractor with a cord holding a plumb-bob. The bob would fall across the instrument straight to the ground, and the gunner would simply read the angle from the scale on the instrument where it was intersected by the cord.

Nonetheless, the task remained of determining how far and how high in the distance a particular target lay. Without this information, the calculations done to determine cannon angle would be moot. The initial solution to this problem was found in the fourteenth century. The cross-staff was an X-shaped device on which the vertical piece slid along the horizontal one, which contained a graduated scale of angles. A soldier would align the top of the vertical piece to match the top of the observed structure and then do the same with the bottom of the structure. He then read the angle off the scale on the horizontal part of the tool. Having determined the angular distance of the city wall or castle tower, he could then adjust the cannon to match the readings on the cross-staff (Burke 1978, 258–259).

Medieval scholars who attempted to explain the motion of projectiles, such as cannon balls, faced a different set of issues than did gunners who simply wanted to know how far their cannon would shoot. The study of projectiles in motion is called the science of kinematics. Like much science in medieval Europe, this traced its roots to the writings of the Greek philosopher Aristotle. Aristotle viewed motion as simply one of the four types of change that an object might experience. He referred to it as local motion, meaning change from one location to another. The remaining three types of change were change in quality, change in size, and the change from life to death. For Aristotle, and for early medieval scholars, numbers and numerical abstraction had no place in the study of nature. For them, there was no basis for a mathematical description of projectiles.

Aristotle explained motion as the result of an item seeking its natural place in the world. For example, stones fell to earth because they were heavy and sought their natural place at the center of the earth, but since they could not get to the center, they settled for the surface of the earth. At the other end of the spectrum, fire was extremely light in weight and sought to be among the heavens. This explained why fire burned up. Aristotle called the fulfillment of an object's natural place "natural motion." There was also "violent motion." This transgression of the natural order occurred when something moved in a way contrary to its nature, such as a stone traveling horizontally through the air. Aristotle argued that for violent motion to occur, the cause of the motion must be in constant contact with the thing moved. In the case of projectile motion, such as a cannon ball or an arrow, Aristotle explained that the air surrounding the cannon ball in flight moved from the front the of ball to the back of it and provided a kind of push. This process lasted until the resistance of the air was too great and the projectile returned to its natural motion and fell to the earth. Expressed in more modern terms: the speed of a body in violent motion is proportional to the force of the initial motion and inversely proportional to the resistance of the surrounding medium. Therefore, Aristotle placed great emphasis on the air in accounting for projectile motion.

Later, medieval scholars attempted to go beyond Aristotle in analyzing projectile motion. The University of Paris and the University of Oxford were centers of cutting-edge research into kinematics. The first task of these academic thinkers was to separate motion from other kinds of change and to make it something unto itself, a separate avenue of study. At Merton College, Oxford, scholars considered the possibility of new concepts such as uniform speed and uniform acceleration, all expressed in words rather than mathematics. The French mathematician Nicolas Oresme (ca. 1320–1382) described motion in terms of a graph where the variables of motions were plotted. However, despite the modern look of Oresme's presentation, he used no numbers but rather relied on the shape of the curve to discus motion.

Medieval thinkers were critical of Aristotle's explanations for violent or projectile motion. If the Greek philosopher was correct, then it should be possible to move cannon balls and arrows simply by agitating the air behind them, which it clearly was not. New explanations argued that the mover (cannon or bowstring) imparted to the projectile a power of motion, known to later medieval writers as *impetus*. The search for this power of motion began with Islamic scholars, who called it *mail*. Ibn Sina (980–1037), known as Avicenna, wrote that *mail* was proportional to the heaviness of an object and that *mail* would remain in a projectile so long as the resistance of the surrounding medium (usually air) was not too great. Both Islamic and western scholars accepted this new explanation of projectile motion. Jean Buridan (ca. 1295–1358), a French scholar who studied at the University of Paris, was the main western proponent of *impetus.* Buridan suggested that the amount of *impetus* in a moving object was directly proportional to the weight of the object and the speed at which it was moving. He also postulated that if no resistance was present, a body in motion would keep its *impetus* forever. This ideal situation, as medieval gunners knew very well, never occurred.

TOWN AND CASTLE DEFENSES

The city wall was the primary form of defense constructed in medieval Europe. Many cites and towns used old Roman or Viking-era walls as their fortification, but, by the thirteenth century and with the advent of continuing urbanization, the limits of these barricades were quickly exceeded. Medieval city walls demarcated the limits and influence of the city. Walls defined a city and were a source of civic pride, offering protection for person and property. The richer the city or town, the more pressing the need for walls, and, what was more, large, imposing walls were often a sign of prosperity. Conversely, less important towns were left unwalled.

While stone walls tend to remain as lasting testaments to their urban origins, most medieval defensive walls were earthen. Walls assembled from dirt and wood continued to be popular into the 1400s. Walls of stone, if they were built at all, were erected last, after a moat and a variety of earth barriers. Construction could take as long as a century to complete but rarely lasted more than several decades. These walls were undertaken as a collective enterprise, with the financing being spread over the entire population. The burden of paying for town walls was, at least in England, alleviated through the imposition of taxes, but not on the inhabitants. Through a special royal decree, first issued in 1220, a town enacted a murage tax on all goods brought into the town to be sold. Thus, the expense of town walls ironically often fell most heavily on outsiders (Kenyon 1990, 183). Most city walls were similar in dimension: they were usually about 2.5 yards thick and 8 to 13 yards tall, with some extreme examples reaching heights of 26 yards. Walls were equipped with walkways and arrow slits for retaliatory assaults.

Generally, all the defensive structures present on castle walls were also present on city walls. Roads leading in and out of the city were secured by imposing gatehouses, with the road passing beneath a heavy wooden door and through a lengthy covered hallway before entering the city proper. Gatehouses also included such civic necessities as tax offices and custom offices. In some urban settings, where a local noble ran the city by collecting taxes in exchange for promises of security, the citadel dominated the landscape. A citadel is a fortress within a walled city that often overlapped the walls of the city. This allowed one to enter the structure without having to enter the city. Like its larger brethren, the castle, the stone structure of the citadel was the focal point of defense for the

City walls outside Florence.

surrounding area. Also, from the citadel the lord exerted influence over those who worked for him.

By the end of the fourteenth century, the arrival of gunpowder weapons changed the shape of city walls, and, by the beginning of the fifteenth century, no formerly impenetrable fortress was secure from the new cannon. Pre-fifteenth-century walls were constructed with flat surfaces that provided too easy a target for cannon. The first modification was the addition of guns on city walls, which required cutting holes (gunports) into the walls for the muzzle of the gun. Other changes meant to answer the arrival of cannon included thickening existing walls with mounds of dirt to absorb the cannon balls. Another solution was to construct new walls some distance from the original city wall and set artillery onto these outer walls. As the historian Kelly DeVries puts it, the intent was to have gun face gun.

Castles by their very nature were more secure than towns or cities. Most castles incorporated walls, called curtains, into the design. Within these walls were *archères,* or arrow slits, often fashioned by leaving a small spaces between stones through which archers could shoot attackers. Like those of towns and cities, castle walls used ashlar stone, prepared rocks that were formed into blocks and made smooth. A curtain was an outer wall that enclosed a castle; the term could refer to a wall that joined two or more defensive castle towers. The towers themselves, known as bastions, extended above the height of the curtain wall to protect it. Later, bastions were angled in response to the arrival of cannon, in an effort to deflect a cannon ball strike. Traditional rounded towers too had to be modified in response to the advent of cannon. Round towers had a blind spot at the front that could not be covered by the defenders. Should an enemy make it into the blind spot and set up artillery, defenders could not adequately defend themselves. Triangular bastions solved the problem by eliminating the obscured area in front of the bastion (DeVries 1992, 263, 267; Lepage 2002, 256–257).

However, before attackers could even contemplate breaching the curtain walls or bastions, they frequently had to pass a moat, a ditch, with or without water, built to keep attackers away from the walls. Defenders on the curtain wall repelled attacks from inside the hoarding. A hoarding was a structure with holes in its floor that extended over the edge of the wall so that the defenders were suspended above the attackers. The intent was to protect the castle; the hoarding permitted bombs, boiling oil, and other deterrents to be dropped upon the invaders. Originally made of wood, later hoardings called *machicolations* were made of stone. Some castles had defensive structures called barbicans, which were located before the main entrance to the castle. They functioned the same way that gatehouses did in the walls surrounding cities. Barbicans could reinforce already secure castles or protect castles where the existing defensive wall was susceptible to attackers at the gate.

7

MEDICINE: ACADEMIC AND FOLK

The practice of medicine in medieval Europe bears little resemblance to the modern experience. Today, patients expect, and likely demand, a doctor who attended university and received a great deal of formal training. Conversely, in the medieval period, people could live their entire lives without ever being treated by a university-trained physician. Indeed, the everyday medical experience of most people was with medical practitioners of various types who had learned their trade through on-the-job training. The sick were more likely to be treated with prayer and bloodletting than with anything close to modern pharmaceuticals, though herbal remedies were not uncommon. Physicians (the designation was almost always reserved for those who had obtained an M.D. from a university) spent much of their time composing philosophical treatises and contemplating anatomy. Rarely did they actually examine patients or personally administer cures.

This chapter outlines the practice of medicine in the medieval era from roughly 500 to 1500. Though the subject matter is broad, the choice of topics is meant to provide a hint of the contemporary theory and practice of medicine in several familiar areas and some not quite so familiar. First, Galen is presented, because the writings of this Roman physician and philosopher became the basis for most, if not all, medical knowledge in much of medieval Europe. Arabic writers on medicine are considered next because it was thanks to their preservation by Muslim scholars that ancient medical writings survived and developed. Surgeons, who today are admired and placed near the top of the physician hierarchy, held the opposite spot in medieval times and obtained much of their knowledge through the treatment of battlefield wounds. Thus, military medicine forms its own section. The beginning of life in childbirth was filled with theory and custom that are considered next. The great epidemic of the fourteenth century, the Black Death, merits close examination. Also, theories of and cures for mental illness receive attention. Medical training, both in villages and at newly establish universities, rounds out the chapter, which concludes with a presentation of the astronomical and astrological medical theories discussed in Italy and central Europe during the late fifteenth and early sixteenth centuries.

GALEN AND HUMORS

The writings of the Roman physician Galen (129–199 c.e.) dominated medieval medical theory and practice. Roy Porter and Andrew Cunningham, historians of medicine, submit that Galen set the medical agenda in Europe from about 200 to 1700. Galen was born into a wealthy family and acquired his medical training through extensive travel and contact with people from faraway lands, including India and Africa. From around 169, Galen served as an imperial physician. He also worked as a physician to the Roman gladiators, from whose treatment he certainly received much firsthand knowledge about the inner workings of the human body.

One of Galen's best-known works was *De Anatomicis Administrationibus* (On Anatomical Procedures). Within its pages, Galen advocated studying medicine and anatomy through observation. During Galen's day, however, dissection of human bodies was prohibited. As a result, his account of human anatomy developed from analogy, specifically the dissection of apes and other animals. For example, to reveal the continuing action of the heart to his contemporaries, many of whom did not believe it, Galen practiced vivisection (the cutting of live tissue) on animals. With the chest cavity of the animal open, it was simple to see the heart in action. Galen's other important treatise was *De Usu Partium* (On the Use of Parts). The chief argument of the book was that all the parts of the human body are there for a purpose and are ideally designed for that purpose. Such a perfection of design revealed to Galen the beauty of nature and its forethought. Later Christian writers embraced this interpretation of anatomy but replaced inanimate nature with God as the designer. After Galen died, the practice of learning from dissection died out and would not revive until the eleventh century, in Salerno. Human dissection was still prohibited, but work was carried out on animals—usually pigs.

Galen's account of the organs began with the stomach and the liver. It was there, Galen believed, that food evaporated into blood. The veins attracted this blood and used it to irrigate the other organs. It is important to note that there was no conception of the circulation of the blood. Some blood made its way to the heart, according to Galen, where the impurities were taken to the lungs. The pure blood then traveled to the right ventricle. From there it leaked through the supposed pores in the septum (the dividing wall between the two lower sections of the heart) and into the left ventricle. This wall is very fibrous, and it was believed that it had holes, because without the notion of circulation, how else could one explain how blood got to the left side of the heart? In the left chamber, the blood became "Vital Spirits." This spirit then traveled through the body to keep it vital. It was the spark of life.

The Galenic focus on bodily fluids, developed from the writings of Hippocrates (460–377 B.C.E.), is the key to understanding medieval physiology. Just as the universe was believed to be made up of the four elements (earth, air, water, and fire), the human body had four vital fluids, known as humors. These were blood, bile (yellow and black), and phlegm. An illness was thought to be the result of an imbalance of these four fluids. For example, too much bile produced fevers and too much phlegm caused epilepsy. Heredity was also acknowledged: children often had the same humor imbalances as their parents. Having the correct balance of the fluids restored health. A physician usually began with diet modification in his attempts to restore the balance, since humors were created from ingested food. This treatment was particular popular after the Black Death, when attention to diet was seen as a potential cure. When this failed, new methods were needed. To get rid of excess bile or phlegm, doctors encouraged vomiting or the evacuation of the bowels. When the doctor suspected that the patient had too much blood, the solution was to drain off the excess blood through bleeding. Every person was believed to have a personal balance of these fluids, called a "complexion." The humor that dominated the complexion determined the personality of the person. For example, blood was associated with laughter and a talent for music; phlegm with sluggishness and dullness; yellow bile with a strong temper and quickness to anger; and black bile with melancholy and depression (Cunningham 1997, 25–27, 29; Porter 1997, 73–74).

ARABIC-ISLAMIC MEDICINE

During the seventh and eighth centuries, Islam became the dominant faith in half of the Byzantine Empire (the eastern part of the former Roman Empire) and much of the modern Middle East. In the early ninth century Arab-Islamic medical learning became established. This was the age of translation, during which many ancient Greek and Latin texts were turned into Arabic. The enterprise was centered in Baghdad. An early figure was Hunayn ibn Ishaq (d. 873), a Christian from southern Iraq who had come to Baghdad in search of writings by the Roman physician Galen. Along with his students, ibn Ishaq translated 129 of Galen's tracts into Arabic. Galen proved a favorite of translators, and hence Galenic medicine dominated Arabic medical theory just as it would medieval European thinking. Arabic writers also composed their own original works based on ancient examples (Porter 1997, 95).

Medieval Arabic medicine reached its zenith in the tenth and eleventh centuries. Ali ibn al-Abbas al-Majusi (d. ca. late 900s) was born in southern Persia, but little else is known about his life. His major treatise was *The Complete Medical Arts,* which considered both the theoretical and the practical sides of medicine. More famous were the writings of Abu Ali al-Husayn ibn Abdallah ibn Sina (980–1037), who is better known to students of the medieval period as Avicenna. Avicenna created the first comprehensive Arabic synthesis of ancient medicine. He produced more than 200 works, including a medical encyclopedia titled *Kitah al-Qanun, Canon,* or *The Medical Code.* It is divided into five books, or sections. Beginning with the notion of elements and humors, it moves on to the action of drugs and the description of diseases before it concludes with an account of compound drugs. In the East, Avicenna became known as the "Galen of Islam." Avicenna's allegiance to Galenic thought is revealed in his comments on humors: "a moist body into which our ailment is transformed." Following in Avicenna's footsteps was another equally famous Muslim scholar, Abu-l-Walid Muhammad ibn Ahmad ibn Muhammad ibn Rushd (d. 1198), known in the west as Averroes. He was a physician and philosopher but is best remembered for his protracted commentaries and translations of Aristotle. Averroes composed a medical reference work titled *The Book of General Principles,* written between 1153 and 1169. It contained seven sections that addressed anatomy, health, symptomology, and drugs, among other topics.

The vast Islamic territories brought their scholars into contact with a variety of herbs and plants that possessed medicinal properties. As a result, Islamic physicians were leaders in the study and practice of pharmaceuticals. One treatise listed more than 3,000 such drugs, 800 derived from plants and a further 145 developed from various minerals. The historian Roy Porter states that "the value of Arab contributions to medicine lies not in their novelty but in the thoroughness with which they preserved and systematized existing knowledge" (Porter 1997, 98–99, 101–102).

SURGERY AND MILITARY MEDICINE

The discipline of surgery and the practice of military medicine are closely related. As the medical historian Nancy G. Siraisi claims, "It is a truism that war is the best school of surgery." Early examples are found in the history of the Byzantine Empire, where seventh-century medical books describe how to remove arrows. In the fourteenth and fifteenth centuries, it was common for a fully complemented European army to include surgeons. The English king Henry V's famous battle at Agincourt (1415) was fought with an army that included surgeons whose job it was to care for the rank and file. However, it was frequently the case that surgeons were not immediately available in the battlefield, a situation that often left soldiers to care for their own wounds and those of their comrades. It was common for soldiers to suture wounds without any preparation. Without question, the level of care provided by these surgeons on the spot was rudimentary at best.

Despite their importance to armies, surgeons sat at the bottom of the medical hierarchy in much of medieval Europe. Only in Italy did surgeons and physicians mix freely. The different educational paths taken by surgeons and physicians is the reason for this segregation. Where a true physician trained at a university and became steeped in the medical philosophy of such ancient authors as Galen, surgeons were, generally speaking, the recipients of on-the-job training. In his *Definition and Objectives of Surgery,* composed around 1267, Theodoric Borgognoni, Bishop of Cervia (1205–1298), advocated that young surgeons visit the operations of more experienced practitioners "and commit them to memory" (Grant 1974, 798). Surgeons knew techniques and basic anatomy but rarely the philosophical theory that defined a physician with an M.D., and this difference kept the two professions separate for centuries.

Before the end of the fifteenth century, when cannon and handheld firearms became more prevalent, military surgery dealt with arrow and swords wounds. In spite of the fact that most surgeons became quite proficient at removing foreign bodies from wounds, battlefield lacerations were more often than not fatal. Surgeons had two options for treating wounds. The traditional method involved stopping the bleeding with a hot iron to cauterize the opening. The second approach concentrated on cleaning and applying ointment but leaving the wound open. This difference led to a controversy regarding the proper treatment of wounds. One school of thought suggested that some pus in the wound was desirable and should even be encouraged because pus was a waste product of the body and should be evacuated, so the wound should be left open. The other school advocated using wine to cleanse the wound before closing it. Prevention of pus was the goal of this second method. Henry of Mondeville (d. ca. 1326), a student of Theodoric and a surgeon in the French army, asserted that "It must be concluded that the treatment on which no pus is formed, in which one avoids it as much as possible, is better, surer, and more healthy than that in which it is produced or provoked" (Grant 1974, 805). At the root of the dispute was the question of whether a dry or a moist environment was best for healing. Both sides found support in the writings of the Roman physician Galen.

Gunpowder added a new dimension to military medicine. Battlefield surgeons loathed the new substance, with one English surgeon calling gunpowder one of the most "diabolical instruments of war." Being shot was bad enough, but the cure might be worse. Giovanni da Vigo, an Italian surgeon who, in 1514, produced the acknowledged authoritative work on the subject of wounds caused by gunpowder weaponry, endorsed the practice of treating such wounds with boiling oil. Da Vigo stated that the wound had to be cauterized because the gunpowder had poisoned the surrounding tissue, and treatment required that boiling oil be poured into the torn flesh with or without removing the projectile. His book was popular, with 40 editions being printed in several languages. For decades this was the common treatment. In 1536, however, the French surgeon Ambroise Paré was also treating gunshot wounds. On one occasion he ran out of oil and had to improvise with a mixture of "the yolke of an egge, oyle of Roses, and Turpentine" to treat the wound. To his amazement, the patients treated with the new mixture healed much faster and were in less pain than those soldiers treated in the established way (DeVries 1990, 134–136; Siraisi 1990, 169–170). A new method for military injuries had been found—one directed at cleaning and care.

CHILDBIRTH

The birth of a child in the medieval period was cause for celebration and joy in an age where death and disease were constant companions. A new life could not escape the harsh realities of the medieval existence, and many children did not live to see a first birthday. Even the process of birthing itself was filled with dangers, and either the mother or the infant, or both, might not live through it. In post-plague Italy, the majority of infant deaths shortly after birth may be attributed to complications arising from premature births. However, in other cases it is simply not known why the child did not survive infancy. Financial status was a major factor in rates of infant mortality. Poorer women were at greater risk during childbirth because of the squalid conditions in which they lived and because their generally inadequate diets resulted in sub-par health. Conversely, wealthier families had specific birthing items such as fancy clothes for the mother to wear during labor and special baby sheets and pillows. Rich women also tended to give birth in cleaner environments. Consider the example of Elizabeth de Bohun, Countess of Hereford and daughter of Edward I of England: elegant baths were arranged (this was a special consideration in an age when bathing was done irregularly at best), and she also had the company of clergy and two monks should she wish to confess anything prior to delivery (Musacchio 1999, 26–27; Woolgar 1999, 98).

Medieval Europeans had an accurate description of female sexual organs, internal and external, from the Greek physician Soranus's *Gynaecia*; however, inaccurate translations ensured that such knowledge was not available to all medical practitioners. By the early 1300s, dissection of female bodies produced the interpretation that the ovaries functioned as did the testicles in men and that their purpose was to dispense a thick liquid that brought women pleasure during intercourse. This view of the ovaries may be traced to the ancient writings of Aristotle and Galen, who both argued that female reproductive organs were inverted male ones. Galen and Aristotle further believed that the male sperm was almost entirely responsible for conception on its own and that the female contribution was very limited. Historians suggest, however, that despite women's theoretically minor role in conception, childbirth itself was virtually a female-only event.

It is a common view that women's health care was the responsibility of female healers or medical practitioners (the title "physician" was generally reserved for university-trained doctors, and hence it is anachronistic to apply the term to women in medieval Europe). The usual designation for these women is "midwife," a catchall name for female healers. The scholar Monica H. Green contends, however, that the term "midwife" did not refer to all female medical practitioners, some of whom often treated patients of both sexes. What is more, she argues that male physicians were also interested in obstetrics and even composed learned texts on the subject. While male physicians commented upon all aspects of women's health, there seems to be no evidence of a man examining female genitals. It is here that a gender difference appears in medicine. Childbirth, which deals almost exclusively with close examination of and contact with a woman's vagina, was the natural realm of women, and thus women ran the show. It was not uncommon, however, for male physicians to be in the room as advisers during childbirth. This was especially true for wealthy women, who would often have a personal physician as part of their household staff (Green 1989, 58, 73–74).

Women attending childbirth. The Pierpont Morgan Library, New York. MS M. 638 f. 19r.

When a woman was about to give birth, the midwife took charge of the situation. At first, the pregnant woman was seated upright on the edge of a bed; toward the end of the fifteenth century, women began to give birth on birthing stools (a chair with the center part of the seat removed). In either case, the mother-to-be reclined on the stool or bed and gave birth in a seated position. The midwife would often use buttered hands to help the baby out into the world or to press on the belly. To encourage a reluctant baby to leave the womb, the mother might try walking and jumping. Some midwives employed charms and amulets placed on the abdomen to help speed the delivery. It was not uncommon for the midwife to believe that the womb wandered throughout the chest cavity even right before delivery. (A wandering womb was thought to be the chief cause of hysteria in women.) The impudent womb had to be coaxed back into the proper position to release its precious cargo. As the historian Rolande Graves describes the process: "Sometimes a piece of clothing worn by the father of the baby at the time of conception was placed between the patient's legs to 'draw' the uterus down." The tools of the midwife included scissors and thread to cut and tie the umbilical cord. She might also bring a vinegar-soaked rag to place in the vagina to halt bleeding (Graves 2001, 67–69).

Caesarean sections had been conducted since antiquity. The term itself is the subject of some historical debate. Legend holds that the Roman dictator Julius Caesar was born by the method. More recent historical studies suggests that the term comes from the Latin word *caedere,* meaning to cut, and some scholars believe that the process was developed during the reign of Julius Caesar and that it is why delivery by incision is called a Caesarean delivery. During Roman times, the surgery was performed only when the mother was dying or already dead, since mothers did not survive early Caesarean sections. This fact would tend to discredit Caesar's own birth as the source for the term, because his mother survived Caesar's birth, something quite unlikely had the dictator been born via the procedure. In medieval Europe, baptism was frequently a key element in the performance of Caesarean sections. Dead children could not be baptized, and unbaptized children could not be buried on church land. What was worse, unbaptized children did not go to Heaven but spent their afterlife in limbo. When the mother seemed unlikely to survive and the baby might be saved, a Caesarean was done to secure the life of the child and also to save its soul. However, should both mother and child die prior to the start of delivery, then both would be permitted a burial on church land. The method of performing Caesareans was part of the curriculum at some medical schools in the eleventh century. The procedure was the subject of contemporary debate, and many physicians refused to perform it on any woman who showed even a hint of life (Graves 2001, 94–95).

BLACK DEATH

Between 1315 and 1317, Europe suffered major famines in which half a million people died. Historians refer to the famine of 1315 as the Great Famine. Those who came of age during this time, it is argued, had weak immune systems and were unable to fight off the epidemic that would soon hit all of Europe, parts of Asia, the Middle East, and North Africa. The Black Death (also called the plague or bubonic plague) originated in Asia and was brought to western Europe aboard Genoese trade ships in 1346. While in the East, these Italian merchants had engaged in local battles in which infected corpses were hurled at opponents in the first recorded instance of biological warfare.

The Black Death was actually two diseases, the bubonic plague, which had a mortality rate of 50 to 60 percent, and pneumonic plague, transmitted when plague victims coughed up blood and other fluids and which had a mortality rate of 90 to 100 percent, with death occurring in less than one day. In the nineteenth century, researchers discovered that the pathogen (the microscopic organism that caused the disease) responsible for the plague was a bacterium called *Yersinia pestis*. It is found in fleas, specifically the *Xenopsylla cheopis* variety carried by rats. The first symptoms of plague were a swelling in the lymph glands in the armpits and groin; some glands attained the shape and size of an egg and others the size of apples. The swellings (known as buboes, hence bubonic plague) then appeared all over the body. Death occurred within three days of the appearance of the first bubo. So many died that new graveyards were constructed, as churchyards could hold no more bodies. Estimates of the total dead range from 19 to 38 million, or between 25 and 50 percent of all the people in Europe. Those who lived in close quarters were most at risk, and the poor and sailors were frequent victims. Italy was the hardest hit nation in Europe: Genoa and Venice suffered losses of 50 to 60 percent of their total populations. The Black Death traveled along trade routes as people attempted to flee the devastation, and traders and merchants unknowingly carried the infection with them. It arrived in England during the summer of 1348 and killed almost half of the population.

Various preventive remedies were tried, such as sniffing amber-scented powders and taking strong herbs that were thought to have cleansing properties. This was done because many scholars believed that the plague was a reappearance of an ancient Greek disease caused by *miasma*, the contamination of air by toxic vapors. As a result, physicians ordered their patients to decrease their intake of air by losing weight; those of a larger size, such as fat men and pregnant women, were thought more at risk. What is more, plague sufferers and those deemed to be carriers who would pollute the air were quarantined from the rest of the community. Sanitation and cleanliness were common prescriptions. As one contemporary account states: "every foul stench is to be eschewed, of stable, stinking fields, ways of streets, and namely of stinking dead" (Horrox 2004, 176). Still other physicians sought cures in different ways; some suggested the benefits of drinking one glass of red wine 15 minutes before bed to fend off the plague.

Other medical scholars thought that one's only hope was a direct appeal to God. This was the opinion of the medical faculty at Paris in 1348. Some saw the plague as a punishment from God and sought to flog themselves and thereby avoid the wrath of God. These people were known as Flagellants, and they wandered throughout Europe whipping themselves and causing hysteria wherever they went. The Flagellants were popular Germany in 1348. Flagellants arrived in England in 1349 and were quickly deported as undesirable aliens. In October 1349, Pope Clement VI condemned Flagellants and urged public officials to crush them. By 1350, the movement was destroyed. Europe had little time to recover from the first visit of the plague between ca. 1348 and 1353; it returned in 1360–1362, 1367–1369, 1373–1375, and 1390–1393 (Benedictow 2004, 9, 14).

MADNESS

Mental illness or madness in medieval Europe was thought to result from either a physical aliment caused by an imbalance in the body or a punishment from God. In the case of the latter, for example, madness could be explained by demonic possession. Such a takeover of human faculties by an agent of Satan could occur only if the victim's soul had previously been weakened by sin or if God Himself enabled the demon to enter the insane person. Why God would allow mental illness was a debated topic among theologians, who concluded that God was giving victims a taste of what awaited them in the afterlife should they not reform sinful practices. Prayer, repentance, and exorcism were the proper treatments. Madness, in this interpretation, was a moral disease that could be cured through righteous behavior and a proper Christian life. To protect the rest of the community from the harmful presence of the affected person, the insane were frequently banished. Other treatments for this kind of supernatural mental illness included whipping or starving the demon or devil out of the affected person (Rawcliffe 1997, 10–11).

When the state of a mad person could not be accounted for by means of some moral transgression, a physical cause was thought to be the answer. Madness was not new in the medieval period; it had been known in Greek and Roman civilization. For these two antique cultures, and continuing into the medieval period, care of the insane was the responsibility of the family and not of society. Ancient physicians, such as Galen, argued that the chief cause of ailments of the mind was an imbalance of bodily humors: blood, phlegm, and yellow and black bile. The most frequently cited culprit was an excess of black bile, which tended to make a patient depressed and melancholic. Galen further suggested that either a shortage or an excess of yellow bile caused mania. Female madness, often referred to as hysteria, was, according to Galen, caused by women's wombs "wandering" throughout their bodies. Marriage would cure this problem and settle an active womb.

Medieval theories on madness closely followed Galen's writings on the subject and divided the ailment into four types: frenzy, mania, melancholy, and fatuity. Each condition came from an imbalance of bodily humors. There was no defined treatment that sought to restore the balance of humors. Bleeding and shocking (tossing the affected person into cold water) were remedies for madness. Milder treatments might include purging or abstaining from certain foods or activities. The insane were often banished from their towns and forced to live in what facilities then existed. Most often the mentally ill resided in monasteries or in special wards of the newly emerging hospitals. For example, in 1247, Saint Mary of Bethlehem was founded to care for the mentally ill in England. The facility eventually became known by the pejorative "Bedlam" as a commentary on the sordid conditions there. Such accommodations, however, were basically warehouses for societal outcasts rather than premodern mental hospitals (Porter 1997, 81–82, 127).

MEDICINE IN THE VILLAGE

Medieval medicine in this period was more art than science. Most of those who practiced medicine did not receive their training at universities. More commonly, they had served an apprenticeship with an established master of the trade. This was not the case, however, in Italy, where lower-level medical professionals like surgeons and apothecaries (today's pharmacists) were admitted to universities. The everyday medical care of people in medieval Europe occurred within their towns or villages and rarely involved a university-trained physician. Indeed, local monks and clerics who knew a little medicine might act as healers. While physicians at medical schools and faculties learned the humor theories of Galen, a host of other medical practitioners relied on common sense, though some similarities between the scholar and the professional did exist. Village practitioners, too, embraced the notion of humor fluids as the key to a healthy body and as the key to diagnosing illness, even if they did not know the full theoretical underpinning of doing so.

An anonymous thirteenth-century writer noted that the two best indicators of health were urine and the pulse. According to him, there were 10 distinct types of pulses that denoted various conditions, including the hour of death. A slowing and failing pulse was a sure sign that death was not far away. By feeling the wrist of the patient, a healer could determine exactly what was wrong simply by noting the number and strength of the pulse beats. Bleeding, done with either a knife or leeches, was a common cure used to restore a balance of bodily fluids because blood contained elements of all the humors and ensured a proper pulse. Urine was a particularly good diagnostic tool because it was the "residue of the humors." There were 20 discernible colors of urine, with each indicating a particular illness or condition. Color, however, told only part of the story; a truly dedicated healer would want more information, information that could come only from handling and even tasting the urine to reach a proper diagnosis. The method was not foolproof, as a fourteenth-century account of a plague victim reveals: "the urine appears fair and shows good digestion, yet the patient is likely to die" (Horrox 1994, 175). Although these examples are somewhat sensational, the most common type of care directed at patients was the recommendation of a particular diet to regulate the creation of specific humors in order to restore balance to the body. Also, parts of various plants were fashioned into assorted drugs. More unusual treatments included the use of a variety of enemas.

Village healers often claimed great powers. In one example, a somewhat mysterious medical man had discovered a method by which a nose that had been severed in a battle or animal encounter and subsequently lost might be grown and reattached. After swearing the patient to secrecy, he proceeded to outline the procedure. He first made a template of the former nose and traced it on the back of the patient's arm near the elbow. Next, he cut a flap of skin in the exact shape of the template to produce a nose-shaped piece of skin that hung from the arm. Then he sutured the flap (still attached to the arm) onto the empty space on the face, binding the arm to the head to prevent tearing. After waiting some 10 days, he cut the flap off the arm and sutured it to the face in such a fashion as to closely resemble the severed nose (Grant 1974, 807–808). This may well be the earliest description of plastic surgery.

Just as the understanding of mental illness in medieval Europe often had religious overtones and cures based in theology, so, too, did village medicine. Some Christian writings taught that illness resulted from divine providence in retribution for some sin or transgression against God. In such a case, no amount of medical knowledge would cure the affected person. Prayer and repentance were the only recourses. From the eighth to the eleventh centuries, the Catholic Church turned sites associated with the lives of saints into shrines and places of healing. Pilgrims to these holy places hoped to earn God's favor and perhaps cure their specific illnesses. St. Luke could be called on to help with a variety of maladies, while St. Roch seemed very effective against the plague and St. Radegund healed ulcers. There was still the possibility of a cure based on natural agents, since God might have used a naturally occurring illness, one that would respond to known cures, to achieve a divine intent. Even so, people still could not ignore saying their prayers (Lindberg 1992, 320).

UNIVERSITY MEDICAL FACULTIES

The earliest organized medical training facility in medieval Europe was founded at Salerno, Italy, during the eleventh century. The historian David C. Lindberg claims that Salerno is perhaps best characterized as a center of medical learning, rather than a proper medical school or medical faculty attached to a university (Lindberg 1992, 326). Its growth was stimulated by translations of medical texts. Around 1071, a former African slave named Constantius Africanus (ca. 1020–1087) arrived in Salerno. Africanus translated many Arabic writings on Galen and other Islamic medical treatises into Latin for the eager Italians. Galen's formerly lost writings were available to western scholars for the first time since the sixth century C.E. Students at Salerno codified the known Greek and Arabic knowledge into a single volume, titled *Articella* (Little Art of Medicine). However, it was not solely the literary arena that made Salerno so influential. Rather than concentrating exclusively on the ancient writings, pupils at Salerno read works that incorporated both ancient medical wisdom and material based on practical experience. To that end, Salerno restored anatomy and the dissection (of animals) as features of medical training.

The twelfth century was the great age of universities, with institutions of higher learning appearing in Paris, Oxford, Cambridge, Montpellier, and Bologna. Affiliated with the universities were medical schools and faculties. When taught at the university, medicine achieved a certain degree of legitimacy by being linked to recognized disciplines of knowledge such as logic, rhetoric, and the other liberal arts. A bachelor's degree in medicine could be earned after seven years of study, and a doctorate in medicine after 10. The curriculum, like that in other medieval university faculties, was based on lectures from authoritative texts, usually works from antiquity. In the case of medicine, the key textbooks from which the professor would read were *Articella* and the Arabic scholar Avicenna's *Canon,* both of which were based on the writings of Galen. As a latecomer to the medieval university, medical studies did not, at first, attract many students. Only 40 students and instructors are known to have studied medicine at Oxford in the fourteenth century. The number increased to only 56 in the fifteenth century. Even the petition to King Henry V in the 1420s that sought to restrict medical practice to university-trained physicians did not drastically increase the number of English medical students at Cambridge and Oxford. At German universities, too, medicine was often the smallest faculty (Siraisi 1990, 56–57).

By the middle of the thirteenth century, the University of Bologna became the center of medical learning in Italy (indeed, all Europe), superseding Salerno, which had shone brightly for a short period before it began to decline. During the period 1419–1434, Bologna graduated 65 M.D.s, by far the largest number of any European university but still, on a per-year basis, quite a small number in comparison to other faculties. After 1300, Bologna began requiring that future physicians attend at least one dissection during the tenure of their studies. Before this time, and continuing at other European institutions, a physician could graduate with an M.D. without ever having seen the inside of an actual body (even an animal body). Dissection of humans was prohibited by the Church fathers, including Tertullian and St. Augustine. The first human dissections in medieval Europe occurred at Bologna by the early 1320s. The practice was supervised by Mondino de' Liuzzi (ca. 1270–1326). Liuzzi taught Galenic anatomy but did it in a way unknown to the Roman physician: the anatomical demonstration. The professor sat in a high chair looking out and down upon the students. During a process that might take three or more days, a surgeon (in medieval Europe, surgeons were quite low in the medical hierarchy) or an assistant would perform the actual dissection and point to the part, or organ, being discussed.

There was, as in other studies, a higher purpose to the study of medicine. Academic medicine was seen as an aspect of natural philosophy, a means to gain knowledge of God and the creation through, in this case, knowledge of anatomy. As the historian Andrew Cunningham describes the medieval motivation, "Man is and contains all things: to study him, his body and soul, is to study God's creation in summary, to study God's highest creation, to study the image of God" (Cunningham 1997, 41). It was more of a philosophical ivory-tower type of discipline, one that dealt with theory, rather than practice. Actual treatment, such as bleeding, was carried out by a surgeon or another with less standing in the medical profession.

MAGICAL MEDICINE: FICINO, PICO, AND PARACELSUS

During the medieval period, it was widely accepted that the celestial realm (the stars and planets) had a direct influence over events in the terrestrial world (the Earth), including medicine. Every person, from the time of birth, was awash in a specific and constant sea of celestial forces: both in the air that the person breathed and in the space through which he walked. For example, the Black Death could be explained by an unusual alignment of Jupiter, Saturn, and Mars, an alignment that, even though it occurred in 1345, was still causing trouble during the period of the plague in 1347–1351. In the case of medicine, the preparation of pharmaceuticals had to be done under optimal astrological conditions so that the curative agent would be most effective (Lindberg 1992, 339).

The recovery of Plato's corpus of dialogues in the late fourteenth century spawned a Platonic revival centered in Italy. Among the concepts receiving renewed consideration was that of a universal soul linking all aspects of the universe. One of the earliest Italian neo-Platonists was Marsilio Ficino (1433–1499). Sponsored by the famed Medici family in Florence, Ficino translated all of Plato's works. Ficino then came to reject the philosophy of Aristotle, which formed the basis of all university learning, as too secular and too concerned with this earthly world, instead of contemplating the heavens, in favor of Plato. Plato's philosophy, which minimized the importance of the material world, appealed to Ficino. In the words of the historian Charles G. Nauert, Jr., Ficino "conceived himself as a physician of troubled and delicate souls, guiding them ... towards a contemplative equilibrium" focused not on everyday events but on the higher level of God (Nauert 1995, 62). Ficino viewed the universe as a hierarchy, with uncreated God at the top and all other creatures following in an unbroken chain in which humanity sat near the center.

Giovanni Pico della Mirandola (1463–1494) was an Italian contemporary of Ficino, and the two shared many intellectual ideas. Pico, too, was a Platonist who believed that an ancient knowledge had once existed and was now lost. Though this forgotten wisdom had once flourished, only remnants currently existed and were spread through the works of many authors. Platonic writings, Pico suggested, held the key to recovering this learning, which revealed the interconnectedness between the universe and humanity, giving those who possessed it the keys to relationships linking all parts of creation to one another and to the stars. Pico stated that this led not only to medicines but also to knowledge of God.

Platonists like Pico and Ficino believed that all the parts of the chain were related and that the influence between them ran downward. If one understood these relationships, argued Ficino, then one could use them to effect things such as medical cures. For example, if stellar conditions caused a particular illness, a remedy similarly would be fashioned from powers held in the stars. In 1489, Ficino published *De Vita Libri Tres* (Three Books on Life) as a medical textbook emphasizing heavenly cures. By directing the therapeutic powers contained within the stars and planets through such devices as talismans, a physician could draw down and control the proper healing power from the sky. As Ficino himself put it, "Ourselves and all things which are around us can, by way of certain preparations, lay claim to celestial things. For these lower things were made by the heavens, are ruled continually by them, and were prepared from up there for celestial things in the first place."

Paracelsus (1482–1546), born in Switzerland as Philippus Avreolus Theophrastus Bombastus von Hohenheim, was the son of a physician. He came to reject the learning and authors of antiquity and, around 1529, took the name Paracelsus, meaning greater than Celsus, a Roman medical writer. Paracelsus would later claim that his education came not from ancient Greek and Roman texts but from everyday experience. This rejection of traditional learning is seen in his choice to write in German rather than in Latin, the recognized language of medieval scholarship. At the heart of Paracelsus's rejection of university medicine was its dependence on pagan authors. For a devout Christian scholar like Paracelsus, God's creation (known as the Book of Nature) was the only medical textbook.

Like Ficino and Pico, Paracelsus saw a universe constructed as one entity where everything was related. Diseases, he believed, had external causes found in the macrocosm of the universe, which affected the microcosm of the human body. Macro and micro were connected and were images of each other. Changes in the macrocosm of the heavens and terrestrial world caused illness in the human body. To find effective cures, one looked to nature and substances that resembled the organ or part causing trouble—a process known to Paracelsians as the doctrine of signatures. For example, if one suffered an earache, a cure might be found in a plant or mineral that looked like an ear. When this did not work, Paracelsus advocated chemical analysis of bodily fluids such as urine. He postulated that the world had been created using the three substances of salt, sulfur, and mercury. These were not the recognizable variety that one could encounter within daily experience; rather, they were idealized forms or aesthetic principles that existed only at the creation. Through alchemy, however, a truly skilled student of the art might be able to re-create one or more of these building blocks of nature. This knowledge also transferred to medicine. In the process of examining the urine through alchemy (by breaking it down into its constituent parts), Paracelsus could determine whether any of the three principle substances was in short supply. A common cure included ingesting mercury to restore the original bodily balance (Wear 1995, 311–316).

8

THE PASSAGE OF TIME: CALENDARS AND CLOCKS

Modern preoccupation with time and particularly the smallest divisions of it—think of sporting events in which the winner is determined by hundredths of seconds—was entirely foreign to medieval existence. So too was the clock in a recognizable form for much of the medieval period. Other instruments marked the passage of time. King Alfred of England (849–899), for example, timed his activities with candles that burned at known rates. Sundials and water clocks were common and reasonably accurate for the level of precision demanded of them. This is understandable because, for a medieval person, the hour was the smallest division of time that held any meaning or mattered in everyday life. However, the hour itself was not the regular unchanging parcel of time that it is today. Hours were not 60 minutes or 3,600 seconds but divisions of the day at which important prayers took place. While the movement of the sun's shadow on a sundial described hours, they were not invariant; depending on the geographical location and the month of the year, the length of the hour varied greatly. Only toward the end of the medieval era did a modern understanding of hours come to be utilized.

In order to see how time played an important role, though one vastly different from its role today, in everyday life, we explore several topics in this chapter. First, astronomy is examined because it was by observing the sky that people first devised a calendar. Having outlined the foundation of calendars, we then discuss the medieval year, with its religious organization. Within the year, several important Christian events, such as Christmas, took place. The most important by far was Easter. The trouble with Easter was that its date fluctuated, and determining when it would occur was a major mathematical and astronomical puzzle for those medieval thinkers who tackled the issue. Next, clocks—water, mechanical, and town—are described in sequence so that the regulating of time may be seen as a process originating in monasteries and culminating in the workday governed by the circular motion of the hands on a clock. Last, calendar reform is described because it was readily acknowledged that the medieval calendar did not corresponded exactly to stellar observations, including the occurrence of the equinoxes. Perhaps more important for the modern world, calendar reform in the early sixteenth century involved Nicholas Copernicus and his innovative depiction of the heavens.

ASTRONOMY

Extending as far back as antiquity, astronomers have always been the keenest among their contemporaries to track time, both in long intervals of years, days, and hours and in smaller intervals involving fractions of hours such as minutes and seconds. Classical astronomy seems to have been focused on studying planetary positions and analyzing solar and lunar motions in order to determine a proper and accurate calendar. The predictable and repeating motions of heavenly bodies allowed for the quantification of time. Not surprisingly, the sun, as the brightest and biggest object in the sky, formed the basis for the medieval calendar. A year became defined as the length of time it took for the sun to return to some point on the horizon. While the position of the sun would yield the basic structure for the medieval year, the waxing and waning of the moon provided the monthly divisions. The motion of the sun (medieval astronomy was based on a moving sun and a stationary earth) at certain times of the year also defined the seasons. When the sun is at the southernmost limit of its annual orbit around the earth, winter begins. Conversely, when the sun is at its northernmost limit in the sky, summer begins (McCluskey 1998, 4–5).

Determining the location of the sun in its travels could be accomplished through the use of a sundial, an instrument that displays the shadow cast by the sun upon a gradated surface. Archaeological evidence suggests that sundials were common at least as early as 2,000 B.C.E. According to the Greek historian Herodotus, the Babylonians invented the device, which was present in the Roman Empire by 292 B.C.E. Using sundials, one could determine the solstices (the shortest and longest days of the year) and the equinoxes. When the shadow was the longest at midday, it was the winter solstice (around December 21). When the shadow was the shortest, it was the summer solstice (around June 21). Between these two events fall two days of equal light and darkness, March 21 and September 23, known as the equinoxes. Sundials also displayed the daily hours. Again,

Babylonian astronomers originated the idea of dividing the daytime (time of light) and the nighttime (time of darkness) each into 12 equal segments; these, added together, would make up a total day. As a result, the face of a sundial was marked off in 12 sections, and one told the time by reading the position of the shadow. The day itself was described as being the period between the time the sun cast its shadow on a particular point on the sundial and the time it fell on the same spot the next day. The length of the shadow cast determined when in the year a particular day occurred, shadows being longer in the winter, when the sun was low in the sky, and shorter in the summer, when the sun in higher in the sky. Throughout the Roman Empire and into the early medieval period, a sundial was the only way to determine time. For most of Europe, knowing the hour (full, top of, bottom) was sufficient for day-to-day activities. Only astronomers wanted a more exact measure of time, and they turned to the position of stars, the movements of which were linked to the position of the sun (McCluskey 1998).

To construct a calendar, medieval astronomers and their ancient predecessors calculated and observed stellar motions. However, the cosmos that they examined was not the same as that seen by modern eyes. Until the twelfth century, the medieval view of the cosmos was based on the writings of Plato—specifically the dialogue of *Timaeus.* In this work, Plato posited a spherical stationary earth at the center of a spherical universe. The sun moved completely through the great sphere of the universe once a year. The universal sphere containing the sun and the fixed stars of the heavens rotated once a day around the stationary earth. Stars were secured to the interior surface of the universal sphere and rotated with the sphere itself, but the sun and planets also moved along the interior of the sphere in a path called the ecliptic, while the sphere itself rotated. The sun traversed the sphere at an inclined angle of 23.5 degrees with respect to a horizontal equator. Equinoxes occurred when the sun passed the equator, and the solstices happened when the sun was at 90 degrees to the equinoxes. Whereas the sun completed its travel along the ecliptic once a year, the faster-moving moon finished the same journey in one month. In this system, days and months were defined by the position of the sun and the moon on the ecliptic (Lindberg 1992, 39, 42, 90–91).

The most influential figure in medieval astronomy was the Greek philosopher Aristotle. Aristotle believed the earth to be a great sphere at the center of the heavens. He further envisioned a universe containing a series of concentric crystal spheres, each of which held a planet. An outer sphere held all the stars, which were called "fixed stars" because they seemed not to alter their relative positions. These clear crystal spheres pressed against each other with no space between them. The order of the heavens was Earth, Moon, Mercury, Venus, Sun, Mars, Jupiter, Saturn, and the fixed stars. This universe contained no void or vacuum. The Aristotelian universe was further divided into two realms: the areas below and above the sphere of the moon. Above the moon was perfection, with planets moving in unending circular motion. What is more, the planets and stars were constructed out of a special fifth element, called quintessence. Everything below the moon, in other words the Earth and all its inhabitants, were composed of the four elements (air, fire, water, earth) and not perfect: they could experience change such as death and old age. For Aristotle, nothing moved on its own; every motion had a cause. With respect to planetary motions, the movement of any planet was determined by the motion of spheres on either side of it. This was also true for the sphere of fixed stars. At the limit of the universe, there was what Aristotle referred to as the unmoved mover. While medieval Christian writers would iden- tify this unmoved mover as God, Christ, or the angels, Aristotle himself was somewhat ambiguous about the exact nature of this ultimate cause of celestial movement (Lindberg 1992, 61–62).

There were other planetary motions that required more complicated explanations. Through observation, it was known that planets seemed to move faster when they were nearer the sun and that occasionally planets seemed to travel backwards before returning to a forward motion (this is known as retrograde motion). If a calendar was to be based on the heav- enly motions, then these peculiar movements required explanation. Near the end of the Hellenistic age, Claudius Ptolemy (fl. 150 c.e.) attempted to explain what was known to occur in the night sky using known astro- nomical theory. Ptolemy agreed with Plato and Aristotle that the circle was the ideal form and that it described the path taken by planets as they moved in the heavens. However, it was the way that Ptolemy used the circle that allowed him to do what Plato and Aristotle could not, namely to predict planetary position and to explain planets' often contradictory movements. Ptolemy combined circles in his explanations. Each planet was carried on a small circle called an epicycle, which rotated around a point on the circumference of the planet's orbit (called a deferent in this usage), at the center of which was the stationary earth. Around the orbit, the epicycle containing the planet could either remain stationary or rotate independent of the larger circle. Through this explanatory device, Ptol- emy accounted for all apparent planetary movements. The planet could go backwards, dip down into the larger circle, or simply follow its nor- mal orbit. To account for the observed tendency of planets to move faster when they are nearer the sun, Ptolemy devised a different description of orbits. The standard view was that each of the planets orbited in a circle, with the earth occupying the center of that orbit. Ptolemy retained the earth at the center of the orbit and a circle as the shape of the orbit. How- ever, he suggested that the planet moved around its orbit with respect to another point, called an equant point, that was not at the center of the universe. The planet still moved in a circle but now moved faster at cer- tain parts of the orbit as was observed. For the medieval period, predating Copernicus, this was the picture of the cosmos that provided the basis for the understanding of time and the calendar: Aristotelian/Platonic descriptions of the heavens, which moved according to Ptolemy's mechanisms.

Western medieval astronomers were not the only ones who used the heavens to determine time. Islamic scholars did so, too, but in a different way. They reckoned time using an astrolabe, an instrument unknown in western Europe until late in the medieval period. In the opinion of the historian David C. Lindberg, the astrolabe was "the most important innovation for medieval astronomy and time-measurement." An astrolabe is a deceptively simple tool. With it, medieval observers could carry out many calculations. The device functioned as follows: the celestial sphere, divided into quadrants and various arcs, is projected onto a metal disc, called a mater, usually made of brass. Over the top of the mater sits a second rotating disc, known as a rete, that is cut away so that the mater remains visible. The rete contains various pointers that indicate the position of specific stars against the scale of the mater. An astronomer then rotates the rete so that the visible night sky is replicated on the surface of the astrolabe. By reading the scale around the circumference of the rete, one can determine the date and time with tremendous accuracy because the specific depiction of the heavens on the astrolabe can occur at only one specific day of the year and at one specific time.

THE MEDIEVAL YEAR AND FEASTS

The foundation for the organizational structure of the medieval calendar came from the Romans. In 46 B.C.E., Julius Caesar introduced a new civil calendar that was based on a solar year of 365.25 days. Unlike previous systems, the Julian calendar did not use the phases of the moon to determine the months. This was a major change that solved the problem of celestial compatibility. The moon goes through its phases every 29 or 30 days, and this traditionally provided the basis for a month. Twelve lunar months equals about 354 days, 11 days shy of a solar year. In order to make the solar year compatible with lunar months, extra days have to be added into some months or an entirely new month of about 11 days needs to be inserted into the calendar. Caesar eliminated this issue by dividing the year arbitrarily into a series of months that were not defined by lunar events. In the Julian calendar, the start of each of the four seasons (determined by the occurrences of the solstices and equinoxes) was assigned an unchanging day: March 25, June 24, September 24, and December 25.

With the rise of Christianity as an intellectual and cultural force in the early medieval period, the Roman calendar became co-opted to reflect the devotional activities of Christians. Medieval Christianity was a religion of rituals that had to occur in a defined order and on specific days. The creation of accurate calendars was more than a scientific enterprise—it was a form of worship. Determining such annual landmarks as equinoxes and solstices was relatively simple, and the Julian calendar had already fixed their occurrence. By placing important Christian feasts on solstices and equinoxes, a secure link between Christ and the sun was established, and the Julian calendar became enshrined as the calendar upon which the dates were set. For example, Christ was thought to be born on December 25, nine months after the Immaculate Conception on March 25 (spring and winter, respectively). Christmas was celebrated on December 25 as early as 336 C.E. Knowledge of astronomy provided allegorical support for early Christians who wished to fix the date of Christ's birth and death: December 25, or thereabouts, was the shortest day of the year, and since Christ was to bring light to a world of darkness, the date was an obvious choice. What is more, March 25 was the date of the vernal equinox and the beginning of spring. Since Christ was to bring renewed life to the world and spring is the time of rebirth, the date was a natural fit. Over time, the calendar was filled with feasts and celebrations, all occurring on a specific Julian date. Indeed, for medieval people, the year became defined by these feasts or similar events and the number of days between them. Complications arose with the so-called movable feasts. These were events that did not occur on the same day every year but moved depending on specific celestial events. Easter, for example, is celebrated on the first Sunday following the first full moon after the spring equinox. Determining when this would happen for years into the future involved specific mathematical and astronomical operations known as computus. The importance of finding the right date for Easter was doubly important because many other holidays (such as Pentecost and Ascension Day, to name only two) were defined by their relationship to Easter. Major observances and holidays took place throughout the entire year, and there was not a single month without an important religious day. The following abbreviated list gives an indication of the religious nature of the medieval calendar (Singman 1999, 219–233).

January: 1, Circumcision of Christ; 6, Epiphany

February: 2, The Purification; 24, St. Matthias the Apostle (Movable feasts: Shrove Tuesday, the day before Ash Wednesday and Ash Wednesday, the Wednesday before the sixth Sunday before Easter, which marks the beginning of Lent.)

March: 25, The Annunciation

April: 25, St. Mark the Evangelist (Movable feasts: Maundy Thursday, the Thursday before Easter; Good Friday, the Friday before Easter; Easter, the first Sunday after the first full moon on or after March 21; if the full moon was on a Sunday, Easter was the next Sunday. Lent ended at Easter.)

May: 1, Sts. Phillip and James; 3, the Discovery of the Cross (Movable feasts: Rogation Sunday, 5 weeks after Easter; Ascension, Thursday after Rogation Sunday; Pentecost, 10 days after Ascension (there often were several days of holidays after Pentecost); Trinity Sunday, one week after Pentecost.)

June: 24, St. John the Baptist; 29, the Death of Peter and Paul the Apostles; 30 the commemoration of Paul the Apostle

July: 22, St. Mary Magdalene; 25, Sts. James the Greater and Christopher

August: 1, St. Peter in Chains; 10, St. Lawrence; 15, the Assumption of the Virgin; 24, St. Bartholomew

September: 8, The Nativity of the Virgin; 14, the Exaltation of the Cross; 21, St. Matthew; 29, St. Michael

October: 28, Sts. Simon and Jude

November: 1, All Saints; 11, St. Martin of Tours; 30, St. Andrew

December: 6, St. Nicholas; 25, the Nativity of Christ (Christmas); 26, St. Stephen; 27, St. John the Evangelist; 28, the Holy Innocents

Dating the year was similarly a Christian enterprise. Counting the year from the birth of Jesus was popular with many calendar makers and authors. However, it was also common in everyday secular society for years to be counted from the reign of the current monarch. January was often used as the start of the New Year, but not always. In England, for example, until around 1100, the year began on Christmas (December 25); the date was shifted to March 25 (the feast of Annunciation) and remained at this date until the mid-1700s.

Days of the week and divisions within them followed the same Christianizing practice found in the rest of the calendar. The Resurrection was deemed to have occurred on a Sunday, so Sunday was sanctified and remembered with the Eucharist. Generally speaking, however, the Catholic Church embraced a Roman division of the day and built the liturgy of daily prayers around it (Dohrn-van Rossum 1996, 29). A day was divided into 12 hours of daylight and 12 of darkness. The hours were traced by the movement of a shadow cast by the sun upon the face of a sundial. Depending on the time of year and geographical location, the actual time for the shadow to traverse from one marked division to the next was not always the same. Longer hours existed in the summer and shorter ones in the winter. However, these 24-hour days affected only scholars and other learned persons such as astronomers. People involved in everyday life knew a different system most often based on the rhythms of agriculture.

Daily hours were an important part of monastic existence because prayers had to be recited at specific times. The New Testament described the occasions to pray as taking place at the third, sixth, and ninth hours of daylight, in addition to sunrise and sunset. Hours became the interval between specific prayers. Smaller units of time were described as the length of specific prayers. Establishment of this ecclesiastical hourly division of the day in medieval Europe may therefore be traced to monastic influences. Every day within the monastery was divided into an unbroken sequence of divine tasks: meditation, reading, labors, meals, and sleep. The sequence of prayers separated the day into seven periods (or hours). These were matins, prime, terce, sext, none, vespers, and compline. Matins came at sunrise, compline came at sunset, and sext was in the middle. This was the idealized version, of course; actual practice varied with location. The standardization of monastic life under St. Benedict (around 530 C.E.) created a need for more precise timekeeping. There were six (later seven) daytime services (lauds, prime, tierce, sext, none, vespers, and compline) and one at night (vigils, and matins at sunrise). Most of these were designated and set in terms of solar hours, leading to the term "canonical hour." With the formation of the Cluniac order, around 910 C.E., the notion of discipline and regulation became firmly set as the heart of the monastic life. As more prayers formed part of the daily routine, noting the time of sunrise, sunset, and noon was no longer sufficient for accurate time reckoning. Liturgical duty demanded that the same prayers be said at the same time every day. Ever more precise timekeeping devices became required (Dohrn-van Rossum 1996, 36, 61).

Non-Christians too organized their day around devotional acts in the medieval period. Jews, for instance, had to pray three times a day: after daybreak, before sunset, and after dark. However, none of these duties required a clock. Only observation of the sun was needed. Islam required five prayers every day: at dawn, just after noon, before sunset, just after sunset, and after dark. Muslims too therefore relied on the position of the sun to determine time of day.

BEDE AND TIME

The Resurrection of Christ is the central event in the Christian calendar. The fact that the rebirth of Jesus occurs at the same time as the rebirth of nature (i.e., spring) adds to the significance of the season. All four Gospels agree that Jesus was crucified on a Friday and rose on a Sunday and that this all took place near the Jewish celebration of Passover. Other than that, there is little agreement as to exactly when these events occurred. It was thought, however, that the crucifixion took place on Friday, March 25, the day of the vernal equinox in the Roman calendar. For early Christian scholars, fixing the date of Easter was important because it permitted a yearly remembrance of Christ's suffering and formed the backbone of Christianity itself. As Christianity grew, it became important to express religious observances, like Easter, in terms of the civil calendar. What was more, early Church fathers were concerned with separating Easter from the Jewish Passover, which is observed on the fourteenth of Nisan, the first month in the Hebrew lunar calendar. The fourteenth day would be the first day of the full moon. Passover had no fixed day on the solar calendar and could occur on any day of the week. Determining a date for Easter different from Passover yet near the spring equinox involved the cycle of equinoxes and full moons. It was a problem of theology and astronomy (McCluskey 1998, 77, 80).

The person mostly closely associated with solving this problem was a monk from Northumbria, the Venerable Bede (673–735). Bede was born 75 years after England's conversion to Christianity and only 50 years after Northumbria (in northeastern England, just south of the Scottish border) became Christian. He grew up in a monastery, after being placed there at age seven. His education was mostly religious, consisting of singing psalms, expressing piety, and engaging in prayer, but Bede also learned Latin and grammar. Bede's exceptional abilities promoted him to the position of deacon at the age of 19 when the rule was that no one younger than 25 except gifted candidates should hold the position. At some point—it is unknown exactly when—Bede became master of education at St. Paul's. In 731, he composed his famed *Ecclesiastical History of the English People* (Brown 1987, 9, 16–19).

However, Bede's scientific legacy is secured through his work in computus, methods for astronomical and calendrical calculations to determine the movable date of Easter. Around 703, Bede wrote *De Temporibus* (On Time). The first nine chapters deal with measurements of time on the basis of what would be called seconds, months, and years. The last six chapters are focused on calculating the future dates of Easter. In 725, he produced a larger work, *De Temporum Ratione* (On the Calculation of Time). This was the most noted and cited work on the subject for centuries. Bede was not alone in his pursuit of Easter. There were nearly 2,500 different versions of computus documents circulating in Europe during the early medieval era, signifying the importance of the topic.

Easter was linked to both the Jewish lunar calendar, because of its association with Passover, and to the Julian calendar, because of its association with specific days of the week. The Council of Nicea, in 325 C.E., ruled that Easter would be celebrated by all Christians on the same day: the first Sunday following the first full moon after the spring equinox. To determine this day for any particular year was not terribly difficult. However, to determine Easter for years to come involved the successful merger of the lunar and the solar cycles and the combination of them with the seven days of the week. As Bede put it, "The time when Easter is ordained to take place is ... redolent with sacred mystery" (Bede 2004, 151). An early problem arose because the Western Christian Church used March 25 on the Julian calendar as the date of the equinox. while the Eastern Church used March 21 from the Alexandrian civil calendar. Before any unity of celebration could occur, the Eastern and Western Churches had to agree on the date of the equinox. Again, the Council of Nicea set the standard by declaring March 21 the earliest limit for Easter. Bede himself initially fixed the date of the equinox as March 25 in earlier works before moving it to March 21 in his mature writings, and this became the standard.

In its most basic terms, the computus question may be summarized by the fact that the solar year is 365.24 days and the lunar month is 29.53 days. There are not an equal number of lunar months in a solar year—an 11-day remainder exists. Thus, the moon is 11 days older than it was on the same day the previous year. After a period of two or three years, the moon is a month out of step with the solar year because there would be 13 full moons, rather than 12. The historian Evelyn Edson states the issue clearly: "If Easter were celebrated according to the lunar calendar alone ... the date would continually recede in terms of the solar calendar and the vernal equinox would quickly be left behind" (Edson 1999, 58). The key to finding a correct formula for the computus was to discern a cycle that meshed solar and lunar years. This cycle revolves around the point at which the sun and moon return to the identical position that they held at some point in the past. Greek astronomers discovered that this cycle took 19 years. However, this does not account for the days of the week, and, since Easter must be observed on a Sunday, it presented a further complication. A 28-year cycle (taking leap years into account), in which the days of the month would return to where they had been, was therefore added to the calculation. Various systems of 84- and 95-year cycles proved unsatisfactory. In 457, Victorius of Aquitaine determined that a cycle of 532 years (the 28-year cycle of days multiplied by the 19-year cycle of solar and lunar years) was the most accurate. After 532 years, the date of Easter would return to what it had been in the first year of the cycle. Bede adopted and codified this system into a comprehensive account. In his words, "When it has completed this total through the sequence of months and days, it immediately returns upon itself and recommences everything pertaining to the course of the Sun and Moon in exactly the same fashion" (Bede 2004, 155). Bede's work concluded with chapters addressing "The Time of Antichrist," "The Day of Judgment," and "The World to Come," which reveals that the concept of time for medieval thinkers was never divorced from religious significance.

WATER CLOCKS

Early medieval Europeans kept track of time with devices that are unfamiliar to modern eyes. Mechanical clocks with their characteristic tick-tock did not exist until relatively recently. Sundials proved very accurate but relied on clear days and had to be placed outdoors. Telling time on cloudy days, at night, or inside buildings required the invention of a different type of clock. The water clock, or *clepsydra*, was the solution.

The ancient Greeks were the first Europeans to use water clocks, although the technology seems to have originated in China and India during a much earlier period. Water clocks arose out of the desire to measure shorter amounts of time than a sundial would allow. Until the end of the fourth century B.C.E., water clocks were used like modern stopwatches— to time things. In the case of speeches, an orator would be allowed to talk for as long as it took a specific amount of water to flow from one container into a receptacle with gradations marked on the interior. Conversely, time could be measured by noting the falling level of water in the reservoir. The Greeks, and the Romans after them, divided their day into 12 divisions known as hours and adjusted their water clocks to release a specific volume of water over the period of an hour described by a sundial. Once the exact amount of water for an hour became set for a particular clock, then smaller fractions of the hour could similarly be marked. Modern experiments reveal that water clocks of this type are accurate to within 15 minutes over 24 modern hours.

During the Hellenistic period, technological advances improved the water clock, making it easier to use. The major addition to the clock was a float placed in the vessel that received the water. As the water level rose, so too did the float, which was attached to a vertical rod. When the float got higher, a pointer at the end of the rod indicated the hour on a scale. Alternatively, the rod attached to the float could have a series of cut-out teeth that engaged a toothed wheel. As the rod rose, the wheel turned to indicate the passing from one hour to the next. Variations on the water clock included an hour hand that turned by a mechanism like the one just described and presented the hour on a stationary clock face. In this version, the face of the clock would look quite modern except that no minute or second hand would be present.

Regardless of the type of water clocks used, several different professions and types of activities relied upon them. Monasteries were early medieval users of water clocks, especially during the often sunless winter months and at night to ensure that specific prayers were said at the appropriate time. Soldiers on guard duty used smaller portable water clocks to time the length of their watches at night. Physicians used water clocks to time the pulse rates of their patients (Dohrn-van Rossum 1996, 23–27; Landels 1979, 32, 36).

A close cousin of the water clock was the sand clock, or, to use its more popular name, the hourglass. It would be natural to assume that, because of their similar design and operation, water clocks and sand clocks developed at the same time. However, the sand clock evolved much later and in a nautical setting. While water clocks have existed since around 1500 B.C.E., the earliest depiction of a sand clock appears in an Italian fresco painted in 1338 or 1339 C.E., though written records mention them about a generation prior to this. The principle behind the sand clock is identical to that of the water clock: the regulated movement of a fluid (in this case fine-grain sand) from one vessel to another is used to measure a predetermined period of time. What made the sand clock so convenient for shipboard use was that the constant up and down of sailing made the water in water clocks unstable and therefore unsuitable for accurate timekeeping. The flow of sand in an hourglass was relatively unaffected by the bobbing of the ship and therefore kept time with a greater degree of accuracy (Balmer 1978).

MECHANICAL CLOCKS

While water clocks and sundials provided accurate timekeeping for much of the medieval period, they had certain drawbacks. Sundials worked only outside and on sunny days. Water clocks were not susceptible to these limitations but tended not to work in the winter months, when the water turned to ice. In northern Europe, where it tends to be both cloudy and bitterly cold during much of the year, these were serious problems. The mechanical clock, with wheels and gears, was the solution to the issue of telling time on all occasions and at all places. Historians do not know exactly when the mechanical clock came into being. The problem is one of language. In the medieval period, the term *horologium* (clock) was rather generic and referred to all timekeeping devices and not specifically to mechanical clocks. As a result, the change to mechanical clocks cannot be seen in the language of the time. Only during the 1300s is there specific mention of mechanical clocks, although the technology likely predates this reference (Landes 1983, 53).

A mechanical clock is a concert of moving parts and weights. Early versions functioned by means of the acceleration of weights tied to a rod by a rope. The rope unwound at a constant rate that could be controlled by means of gears. This was especially true of clocks mounted high in towers. The rotating rod indicated the passing of the hours, but not exactly how one might expect. These first mechanical clocks sounded a bell or another audible signal at the hour and very likely did not rely upon a dial or hour hand (the minute hand being a much later invention) to show the time of day as did the water clock. This method of signaling changed the name of these mechanical clocks. Eventually, the word *horologium* was replaced by a version of the Latin word for bell, *cloche,* and later this was shortened to the more familiar *cloak.* Such a design worked wonderfully well in tall buildings where the weight could travel a long enough vertical distance. For less substantial rooms, a different system was required.

The technological innovation that made the mechanical clock ubiquitous was the "escapement," but the basic principle remained the same. In this system, gravity-driven weights still rotated on an axle, but the circular motion of the axle was regulated by means of an escapement device. An escapement functions as follows: a metal wheel with teeth links to the rotating axle through its corresponding single tooth, called a verge; the wheel then blocks and releases the verge to govern both the rotational motion of the axle and the descent of the weights. An escapement continually halts the toothed wheel and then permits its motion to continue, turning the axle attached to clock hands or a dial around a stationary indicator. This occurs at regular intervals and allows the mechanism to run independently for a lengthy interval. To reset the clock, one simply winds up the weights. The noise of the clock with its regularized hit-and-release operation is the source of the familiar tick-tock sound. By the 1400s, escapement clocks were the norm.

Now that time had been mechanized, a demand arose for even smaller personal clocks, what for modern users would be known as pocket watches. Clock ownership became a mark of status for kings, princes, and local political strongmen. Smaller clocks, however, could not rely on

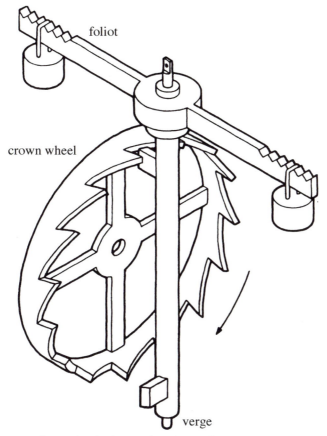

Mechanical clock mechanism. Illustration by David Penny—Copyright.

weight as the force of motion. A different form of drive was needed before clocks could be miniaturized. Rather than employing the rotational power generated by weights, smaller clocks, or watches, used the power of an unwinding coiled spring to turn the indicating dials or hands. There was one problem that had to be overcome before a spring could be effective inside a watch: springs lose power as they unwind. Clockmakers connected the spring to the inside of a barrel (known as a *fusee*) and put a chain around the outside of the barrel. As the spring unwound, the chain moved the main drive shaft of the clock. The key to the regular movement was the tapered *fusee,* which allowed the spring to do easier work at the end by starting the chain on the largest part of the *fusee* and progressing to the smallest. (Think of bicycle gears.) Spring-driven watches first appeared around 1430, and carrying time became the fashion (Bruton 1979, 7–10; Landes 1983, 48–50).

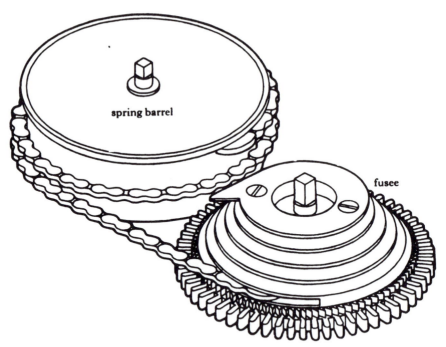

spring barrel

fusee

Pocket watch mechanism. Illustration by David Penny—Copyright.

TOWN CLOCKS

The introduction of large central clocks in towns and cities during the fourteenth century was both a social and a technological issue. In the first place, the cost involved was substantial, easily equal to millions of dollars in modern terms. The financial burden fell to the townspeople or city dwellers. For example, Montpellier, a city in France, raised the salt tax to pay for its clock. A clock was also a prestigious project that brought fame to a city. Clocks were built in communal towers so that they might be admired by all. Competition fueled by civic pride dictated that cities wanted larger, more elaborate clocks than their neighbors had. Employing the same basic weights and wheels as other mechanical clocks, these public-display items often incorporated elaborate techniques to signal the hours, such as a procession of miniature saints ringing bells, roosters crowing at dawn, and rotating disks featuring lessons in astronomy. Italian cities were among the first in Europe to construct large public clocks. Milan claimed a clock in 1336, followed by Padua in 1344, Genoa in 1353, and Florence, also in 1353. By the early fifteenth century, northern European cities were getting in on the act. A 1425 chronicle from the German city of Mageburg stated that the intention to build a new town clock was "for the honor of the city and the utility and comfort of the citizens" (Dohrn-van Rossum 1996, 146).

Tower clock in Piazza Del Campo in Siena, Italy.

Town clocks indicated all 24 hours either through bells or by hands on the clock face. All that ringing wore out the bells, so some Italian cities switched to two segments of 12 hours rather than counting from 1 to 24. However, this created a problem: the clocks measured hours of invariant length. Since the clock turned at a constant rate, the changing length of the hours based on the movement of the sun through the horizon, or as determined by a sequence of prayers, could not be replicated. The new mechanical clocks would reflect an artificial hour: the solar day divided into 24 equal segments regardless of the time of year. This created a contrast between the natural daily routine of the peasant farmer, who timed his or her day by the specific order of the tasks of agriculture, and the invented time of the cities and towns. The sun governed farm time, while time in cities was the property of technology. Performing a job at the behest of a mechanical contrivance became the norm for workers from the fifteenth century on. Bells attached to clocks sounded at the start of the workday, for meal breaks, at the end of work, at the closing of city gates, at the start of market, and to mark many other daily events. The new unnatural time created some suspicion on the part of townspeople. For example, how did workers know that time was being run out fairly? Were employers trying to stretch the day by not ringing bells? When bells were later run at the hour, half-hour, and even quarter-hour, the problem died down. Clock hands further restricted employers' ability to stretch out the workday because everyone could look at the hour for himself or herself.

CALENDAR REFORM AND COPERNICUS

The medieval calendar of 12 months, based on the Roman calendar, had its origins in the solar year, the length of time it took the sun to return to a precise previous position in the sky and on the phases of the moon. Replicating the movement of the sun in an ordered system of days and months often proved quite difficult. This was further complicated by the fact that lunar months did not fit evenly into the solar year. It soon became apparent that observations in the sky with respect to such events as solstices and equinoxes did not match the calendar predictions of them. Calendar reform was the answer.

During the Roman Republic, the calendar contained 12 months: *Ianuarius, Februarius, Martius, Aprilis, Maius, Iunius, Quinctilis, Sextilis, September, October, November,* and *December.* The months *Martius, Maius, Quinctilis,* and *October* had 31 days. *Februarius* had 28, and all the rest had 30, for a total of 355 days in the Roman year. Once in a while, a group of priests called *pontifices* would shorten *Februarius* to around 23 days and insert an extra month *(intercalarius)* of 27 days to give a year that was 377 to 378 days long. (The purpose was to account for a solar year of 365.25 days, which currently is taken care of with a leap year every four years.) This practice was not always followed; for example, it was not always remembered during times of war when things like survival were more important that inserting an *intercalarius* into the calendar. Eventually, the Roman calendar fell badly out of sync with the solar cycle.

This happened during the Gallic War (58–51 B.C.E.) because an *intercalarius* was included only during the years that Julius Caesar spent *Februarius* in Rome. To get back in line with the solar year, Caesar reformed the Roman calendar by ordering that 46 B.C.E. would have 445 days and that *Ianuarius, Sextilis,* and *December* would have 31 days, while *Aprilis, Iunius, September,* and *November* would have 30 days. In a leap year, which was supposed to occur once every four years, the day *Februarius* 24 would be counted twice, but confusion over what Caesar had written led to a leap year every third year. Relying on information supplied by his astronomers, Caesar fixed the length of the year at 365 days and six hours, or 365.25 days. However, this calculation was not entirely accurate. The additional time over 365 days is 5 hours, 48 minutes, 46 seconds, or 11 minutes 14 seconds less than Caesar's year. While this might seem a trivial difference, over the course of decades and centuries it really added up. Later changes to the Roman calendar are more recognizable. In 44 B.C.E., *Quinctilis* was renamed *Iulius* (July) to honor Julius Caesar, who had been assassinated that same year, and in 8 B.C.E., *Sextilis* was renamed *Augustus* (August) to honor the famed emperor.

Scholars in the medieval period noted inconsistencies among the calendar, the sun, and the moon. Because of the 11-minute variance, the calendar was off by three days after 400 years. In the thirteenth century, John Holiwood suggested deleting one leap day every 288 years. There were calls for a papal inquiry to determine the correct date of the equinoxes. Both Robert Grosseteste and Roger Bacon worked on the problem. In 1437, Pope Eugenius IV considered a proposition to delete the last seven days from May 1439, and later he deliberated over a suggestion to delay eliminating October 21–27 until 1740. The problem continued unabated. The Fifth Lateran Council (general meeting of Catholic Church officials), which took place from 1512 to 1517, established a committee to reform the Julian calendar, which, by the sixteenth century, was 10 days out of sync with stellar observations. In 1578, Pope Gregory XIII set forth to reform the calendar, and the results of his study were set down in the papal bull *Inter gravissimas* (February 24, 1582), which declared that in that year October 4 would be followed by October 15. The purpose was to move the vernal equinox (so critical in Easter calculations) from its current date of March 11 (because of the accumulated error of the Julian calendar) to the more proper date of March 21. Not all countries accepted the new Gregorian calendar. Protestant England resisted the change and would be 11 days out of sync with the Continent until 1752, when it finally adopted the new calendar—albeit some 200 years after the fact (Blackburn and Holford-Strevens 1999). This difference between England and the Continent continues to affect modern historians when they attempt to date correspondence sent from England to the rest of Europe and vice versa.

Perhaps the most famous person involved in calendar reform in the sixteenth century was the Polish astronomer Nicholas Copernicus (1473–1543). Officials at the Fifth Lateran Council invited scholars to participate in reforming the Julian calendar and in 1515 asked Copernicus to join the endeavor. In a letter, now lost, Copernicus declined the invitation and explained that in his opinion existing theories of astronomy could not explain or agree on the length of the year and that a reform of astronomy needed to precede any attempt at calendar reform. As he explained to Pope Paul III, "mathematicians are so unsure of the movement of the Sun and Moon that they cannot explain or observe the constant length of the seasonal year" (Kuhn 1957, 125–126). Copernicus remained committed to calendar reform but set out first to re-evaluate contemporary astronomy.

In the sixteenth century, astronomy was based on the theories of Aristotle (a stationary Earth at the center of the universe with all other planets and the sun orbiting it) and the mathematics of Ptolemy (planetary motions described in terms of epicycles and deferents). Copernicus turned this picture upside down in his famed book *De Revolutionibus Orbium Coelestium* (On the Revolution of the Heavenly Spheres), published in 1543. The key conclusions in *De Revolutionibus* are: (1) that the spheres of the planets revolve around the sun, which is near the center of the universe; (2) the motion of the sun is really the motion of the Earth, which moves like any other planet, and there is no ecliptic; it is the Earth that is inclined at 23.5 degrees; and (3) the Earth rotates daily upon its axis, while the heavens remain unchanged. The new order of the planets was Sun, Mercury, Venus, Earth, Mars, Jupiter, Saturn, and the stars. What this organization of stellar bodies also did was to explain retrograde motion in terms of an optical illusion viewed from the Earth orbiting the sun within the larger orbits of more distant planets. For Copernicus, his system was more acceptable than thinking that the entire universe turned around the Earth everyday. In the preface to the book, which was dedicated to Pope Paul III, Copernicus stated his hope that *De Revolutionibus* would aid in calendar reform. It would prove useful in the enterprise, but in a mediated form. The Gregorian calendar used a new set of stellar tables for its creation: Erasmus Reinhold's *Prutenic Tables*. As his source of stellar computations, Reinhold relied on Copernicus's *De Revolutionibus*.

9

SCIENCE AND RELIGION

Consideration of the relationship between medieval science and religion (if we may actually separate them in any meaningful way) is fraught with difficulty that is not entirely explained by the immense distance between the historian and the medieval period. Present-day interpretations of both science and religion almost always color the analysis. This is not surprising in that scholars often try to understand the past in terms that are relevant to modern readers and that resonate with their experiences. When the topics are two of the most politically and ideologically charged in historical inquiry, the result is certain to be debated with a passion not often seen in other forms of scholarship.

The first thing to recognize about medieval science is that it was conducted within an environment that was profoundly concerned about God and religion. This is not to say that religion, specifically Roman Catholicism, was an oppressive force that stifled intellectual inquiry and shut down all science with which it did not agree. Rather, a shared Catholic faith provided medieval thinkers a framework within which to understand the world around them. Nature was seen as God's handiwork and was studied with this in mind. While some processes might seem to operate by nature alone, medieval scholars believed that this was simply the mechanism through which an active God chose to do His work. Also, there was no blanket Catholic response to science. Where St. Augustine, writing in the fifth century, saw science solely as a handmaiden to religion and theology, St. Thomas Aquinas, who lived and wrote in the thirteenth century, some 800 years after Augustine, viewed science as complementing theology in the service of God. Men who would have considered themselves theologians first and foremost did some of the era's best works on optics and scientific method. What to modern eyes looks like science would have been simply one aspect of a very diverse learned activity, but to the medieval mind part of a coherent whole. The term "natural philosophy" was used by contemporaries to identify the activity. Medieval philosophy was divided into various categories, of which speculative philosophy was one. Speculative philosophy was further divided into natural philosophy, concerned with the natures of material things; theoretical philosophy, concerned with abstract quantity; and divine philosophy, concerned with the nature of God. Theoretical philosophy was made up of four disciplines—arithmetic, music, geometry, and astronomy—called the *quadrivium,* which was combined with the *trivium*—grammar, rhetoric and logic—to form the seven liberal arts taught at universities. This acknowledgement is key because the modern understanding of a separation of science and religion is entirely inappropriate when looking at them in the medieval period. Knowing that the worldview of medieval people was not the same as ours leads to an acknowledgment that what was done in studying nature in medieval Europe was not the same thing that modern people would recognize as science, with its goal of dispassionate objective consideration of the world done perhaps while wearing a white lab coat.

As a means to create better appreciation of the multifaceted approaches to this fascinating field of study, it is quite useful to consider what past historians have written on the topic. In other words, what follows is a brief history of the history of science and religion. Popular visions of the engagement between science and religion as one of combative battle may be traced to two books written in the nineteenth century. John William Draper wrote *History of the Conflict between Religion and Science* in 1874. In this book, Draper takes the position that religion was always an obstacle to the march of scientific progress. Twenty-two years after Draper, in 1896, Andrew Dickson White produced *A History of Warfare of Science with Theology in Christendom.* White characterized religion as constantly interfering with science. Since history is never produced in a vacuum and always bears the marks of the historian and culture that produced it, a brief consideration of the environment in which Draper and White wrote sheds some light on their viewpoint. Both men composed their books in the shadow of the reception accorded Darwin's theory of evolution and of the political and religious backlash that it generated. Living during this time affected how White and Draper read backwards into the history of science: every episode of conflict was a precursor to the debates in which they were living. What is more, White wrote while his attempts to further integrate science into the curriculum at Cornell University were being challenged by religiously minded opponents. Despite the more than 100 years that have passed since these works appeared, their view of science and religion as two separate and often combative aspects of human culture remain firmly set in the popular imagination (Osler 1997). More recent historical writing is producing a different and somewhat calmer picture. Taking England as the center of their studies, historians such as Robert S. Merton and Charles Webster, among others, have suggested that Protestantism was conducive to science and actually encouraged its practice. Further studies have also revealed that Catholicism too should be seen as part of this same pattern, especially in the medieval period. The strongest proponents of this view are Andrew Cunningham and Roger French, whose books and articles challenge the persistent view offered by Draper and White.

This chapter, then, offers a picture of medieval science and religion that attempts not to separate the two but rather to show how they were bound in a seamless intellectual activity. It goes without saying that much of the material discussed in this chapter applied only to a very select group of scholars in the medieval era and held very little interest to everyday people. In spite of the elite nature of the topic that follows, its importance for the shaping of medieval Europe (and Europe for centuries after the fact) is difficult to overemphasize. Cistercian monks, in spite their desire to abandon the world and live a life devoted to God, were some of the best industrialists in the medieval period, and a sample of their mechanistic production is presented first. The writings of St. Augustine, Bishop of Hippo, formed the intellectual agenda for many medieval thinkers, and a consideration of his thought rightly finds a place next. Two of the most important figures after Augustine were St. Thomas Aquinas and Roger Bacon, who merit their own sections in this chapter. Translations of Aristotle's Greek philosophy into Latin, which was readable in Europe, shaped the understanding of the world for centuries. The history of this translation process comes next. The introduction of Aristotelian philosophy into the university curriculum created a new form of scholarship known as scholasticism and led the Bishop of Paris to halt what he viewed as the too quick spread of Aristotelian ideas at the University of Paris in 1277. Both of these episodes are presented here. Since the heavens declared the glory of God for medieval people, knowing the workings and potential effects of the sky were of great importance. The chapter therefore ends with a discussion of astrology and portents (omens or warnings) in the sky.

CISTERCIAN MONASTERIES

The intermixing of religious duties with science and technological knowledge is readily seen in the example of the Cistercian monks and the industrial nature of their monasteries. Cistercians were important promoters of technology. Robert, Abbot of Molesmes, founded the order in 1098 when he and a few others who were dissatisfied with what they viewed as a lack of dedication in their existing monastery settled in the forests

around Burgundy (in east-central France) to live according to the monastic rule of Saint Benedict (c. 480–c. 547), away from the rest of society.
This meant that they chose to reject worldly wealth and to live in poverty
and isolation. The golden age of Cistercians is said to have dawned with
the arrival, in 1112, of St. Bernard, who reinforced the ideal of work and
physical labor for the brethren. Cistercians were to exist by the fruits of
their own labor, and each day consisted of a series of regimented tasks.
However, as the historian Jean Gimpel states, the Cistercian order "soon
found itself in a controversial position … for the economic consequences
of these ideals. In their sincere desire to flee the worldly and commercial
life of the cities the Cistercians went to live in areas 'remote from habita-
tion of man.' But by attempting to become independent of the outside
world, they created an economic empire based on a highly centralized
administration and on up-to-date technological expertise" (Gimpel 1976,
46–47).

While some English Cistercians raised sheep specifically to export wool
that was said to be of the best quality, the original Cistercian house near
the city of Cîteaux produced wine that became renowned for its taste
and quality. The sale of wine to the outside community reflects the rise
in living standards in the twelfth and thirteenth centuries among society.
Aside from this market demand, wine was crucial for the Communion
ceremony. Until the thirteenth century, Communion of both kinds (bread
and wine) was taken by all Christians, not only priests, as would be the
case in future years. What is more, St. Benedict permitted his monks to
drink wine (albeit in moderation).

There were more than 740 Cistercian monasteries in the twelfth century, and surviving documentation reveals that practically all of them
possessed a waterwheel. This technology was put to a variety of uses
depending on the particular location. Cistercians outside Provence, in
southeast France, used the power of the waterwheel to operate a grinding
stone to produce olive oil. In Fontenay, just east of Paris, waterwheels powered trip-hammers to forge metal. From around 1250 and into the 1600s,
Cistercians were at the forefront of iron production in the Champagne
region of France. Although by the fourteenth century Cistercian monasteries were on the decline, their embracing of what was then the pinnacle
of modern technology suggests widespread adoption of machines within
all strata of medieval society. Gimpel argues that "the discipline imposed
by Saint Bernard on his monks—the rigid timetable, the impossibility
of deviating from the Rule … brings to mind the work regulations that
Henry Ford imposed on his assembly lines."

ST. AUGUSTINE

The importance of St. Augustine (354–430 c.e.) to medieval thought is
hard to overemphasize. One historian states that Augustine "is the single

most important of the Latin church fathers." As the philosopher Anthony Kenny has put it, after his conversion to Christianity Augustine "developed, in a number of massive treatises, a synthesis of Jewish, Greek, and Christian ideas that was to provide the backdrop of the next millennium of Western philosophical thought." Augustine was born in modern Algeria to a non-Christian family. Many of his works, such as *Confessions* (397), an autobiography of Augustine's early life until his conversion in 387, were theological in nature. In 396, Augustine became Bishop of Hippo, a city in North Africa. His most famous book, written over the period of 413 to 426, is *The City of God*, undertaken as a response to those citizens of Rome who blamed that city's Christians, who had forsaken Roman gods, for the sacking of the city by Visigoth invaders in 410.

In his youth, Augustine followed Manicheism, a heresy that believed matter and spirit were created by individual Gods, with the former being evil and the latter being good. Even after his conversion to Catholicism, Augustine was still influenced by this dichotomy. His attempts to reconcile the world of matter around him and the world of spirits inhabited by angels and God were satisfied in Neoplatonic (that is, derived from Plato, by scholars who wrote after Plato) thought. Although some original Plato (his *Timaeus*) was known in the early medieval period prior to the recovery of the entire Platonic corpus during the Renaissance, it was through Christian theologians, chiefly Augustine, that this Greek philosophy was disseminated among the learned of Europe. Indeed, early Church philosophy might easily be called Platonic. For Plato, the world was at best a copy of the perfect immutable world of spirits. However, copy though the material world might be, contemplation of it did provide insight into the immaterial realm of perfection. Augustine still viewed the material world as corrupt and imperfect, but study of it (what we might call science) led to knowledge about God and Heaven. So long as this hierarchy was maintained, Augustine and later Catholic officials argued, there was nothing subversive or godless about doing science. Problems arose when natural philosophers studied the world for their own satisfaction or as an enterprise unto itself rather than as a means to know God and declare His glory. It must be remembered, however, as the historian John M. Rist points out, that Augustine would not have recognized a distinction or separation between theology and natural philosophy. They were different aspects of the same project, which was to better know God (Rist 1994, 5).

Thus, Augustine viewed science (by which he would have meant Latin interpretations and commentaries on Greek philosophers—chiefly Plato and later Aristotle) as a "handmaiden" of religion. That is, knowledge of the world gained through science was intended to support and aid the cause of Christianity and was never done as something in its own right. Theology was the superior form of knowledge that science would complement but never supersede. What was more, Augustine argued that Christian thinkers must learn about the natural world lest they seem foolish beside their pagan contemporaries. While not all subsequent medieval thinkers accepted Augustine's view of the relationship between science and religion, his position directed learned discussions on the topic for generations (Lindberg 1995, 71–72).

ARABIC TRANSLATIONS OF ARISTOTLE

By the seventh century, Islam dominated eastern Europe. As Arabic-speaking adherents of Muhammad expanded into the territory of the Byzantine Empire, they came into contact with the wonders and unknown riches of Greek philosophy. In the eighth century, the Abbasid dynasty (ruled from 750–1258) moved the capital of the Islamic Empire to Baghdad from its former home in Damascus. At Baghdad, in the so-called House of Wisdom, the Abbasids sponsored the translation of Greek learning into Arabic. The move to Baghdad was also important in that two groups of outcast Christians (both heretical) already called Baghdad home. These Christians had much experience turning out Arabic translations of Greek writings and could serve as instructors to their new Islamic neighbors. Medieval Islamic scholars, unlike their Christian counterparts, believed that learning from other cultures might have merit and could possibly improve their own. What was more, Islamic scholars possessed the ability to manufacture cheap paper made out of pulp and could therefore produce relatively inexpensive copies of Arabic translations of Greek authors.

Almost all the known writings of Aristotle, Plato, Euclid, and Galen were translated into Arabic and read with great enthusiasm, first in Baghdad and later in the rest of the Islamic Empire, including parts of Spain, especially at Toledo, a major center of Islamic scholarship. Arabic translators did not satisfy themselves with simply accuracy. These scholars often added commentary and explanatory notes to the original Greek philosophy. After gaining much attention, the production of translations came to an end around the tenth century.

Avicenna (Ibn Sina) (980–1037) has been called the "greatest of all Muslim philosophers." Known in his own day as an accomplished physician, Avicenna is regarded by historians as one of the dominant medieval sources that preserved Aristotelian philosophy. In addition, Avicenna composed commentaries and glosses on Aristotle's natural philosophy. His *magnum opus* was *Metaphysics,* an Aristotelian treatise of 10 books. When Avicenna's Arabic writings were translated into Latin, around 1150, they had a major impact on western European scholarship. The other Muslim scholar credited with the preservation and development of Aristotle's writings prior to the thirteenth century was Averroes (Ibn Rushd) (1126–1198). Averroes came from a family of judges, yet he practiced medicine. In 1168, the caliph Abu Yakub ordered Averroes to compose a summary of all Aristotle's works. During this task and his remaining years, Averroes wrote 38 commentaries on Aristotle, which would secure his reputation among medieval European scholars as "the commentator." Indeed, while others may have translated more Aristotle than did Averroes, no one was more influential when it came to interpreting those translations (Kenny 2005, 37–38; 47–49).

Not all translators and preservers of Aristotle were Arabic. Boethius was a Roman senator who made Latin translations of Aristotle during the period 510–522 c.e. He also composed commentaries on those writings of Aristotle that he did translate. Boethius had wanted to translate all of Plato and Aristotle but fell short of his goal. By around 1100, Boethius's translations of Aristotle's works on logic were the only ones available in Latin for medieval scholars to study.

During the eleventh and thirteenth centuries, the bulk of the Arabic translation of Greek philosophy was translated once more into Latin. It was during the crusades that western scholars came into contact with the Arabic/Greek corpus and saw that works that had been lost in the West for centuries thrived in the Islamic East. Once discovered, Europeans could not get enough of these formerly unknown or lost Greek masterpieces. Christians and Muslims had close contact with each other in Spain, Sicily, and southern Italy. It was at these locations that the Latin translation enterprise really blossomed. The most important of the three, however, was Spain, which had been under Muslim rule since the eighth century. Many of its residents were proficient in both Latin and Arabic. Latin scholars from all over Europe traveled to Spain in order to learn Arabic and then to translate the works of the ancient Greeks. When Christian forces conquered Spain in the eleventh and twelfth centuries, an even greater wave of Latin scholars descended on Spain. The most famous of the early translators was Gerard of Cremona (ca. 1114–1187). Gerard spent between 30 and 40 years in Spain at Toledo, where, by the end of his life, he had produced between 70 and 80 Latin versions of Greek works.

ST. THOMAS AQUINAS

Thomas Aquinas has been called "the central figure in Dominican natural philosophy," if not all of medieval natural philosophy. He is often credited with the "reconciliation of Aristotelian philosophy and Christianity." Born into an Italian noble family in 1225, Thomas Aquinas began his education as a five-year-old child entrusted to Benedictine monks. His studies continued at the University of Naples, where he learned Aristotle's logic and natural philosophy. Around 1244, much to the dismay of his family, Aquinas joined the Dominican order of friars and set out for the University of Paris. The Dominican order of friars had been founded by St. Dominic in 1218 after he had seen firsthand both the work of heretical preachers and an ineffectual Catholic response. Dominic envisioned his order as a preaching one that would rely on learned followers to spread the word of God. While at Paris, Aquinas met and was befriended by the famed Dominican scholar Albert the Great, who would serve as his mentor. In 1256, Aquinas earned a master of arts degree in theology. With Dominican assistance (Dominicans administered two of the chairs in theology), Aquinas obtained a chair in theology at the University of Paris. He gave up the position in 1259 and by 1261 was affiliated with the papal court, in addition to mingling with diplomats and government officials throughout Italy. During this period, in 1265, Aquinas wrote his famed *Summa contra Gentiles* (Summary against Unbelievers). The book demonstrated points of Aristotelian philosophy that might be shared by Christian, Jew, and Muslim so that this agreement might be used to convert adherents of the latter two groups to Christianity. It was an attempt to use Aristotle in the cause of a Christian mission.

In combating heretics, Aquinas believed that one should use the weapons of an enemy against it. So, when non-Christians used ancient philosophy against attempts at conversion, Aquinas used Aristotelian philosophy as a complement to biblical arguments. Aristotelian writings revealed a universe where everything had a purpose and a place. For Aquinas, this was a form of revelation that demonstrated divine purpose and planning in the world. In his arguments against non-Christians, Aquinas employed Aristotle's *De Caelo et Mundo* (Concerning Heaven and Earth) and *Meteorologica* (Meteorology). Not only non-Christians were Aquinas's targets. A rogue group of Christians called Cathars who believed in two Gods, one good and the other evil, also obtained support for some of their positions from ancient philosophers. Cathars taught that the material world was the product of the evil God and held nothing of merit. Aquinas countered that the material world reveals the existence of God, the creator of the heavens and Earth. The historians Roger French and Andrew Cunningham argue that the replies given to Cathars by scholars, like Aquinas, who used Aristotle in rebuking the heretics marks the birth of medieval natural philosophy (French and Cunningham 1996, 185–195). Not all historians agree with this assessment, and the position remains somewhat controversial.

Aquinas modified Aristotle when it seemed necessary for his Christian mission. For example, where Aristotle ascribed the ultimate cause of planetary motion to a somewhat ambiguous unmoved mover, Aquinas used God. Aristotle had advanced the notion of an eternal world; Aquinas turned this into a kind of philosophical existence that was compatible with the account in Genesis of creation out of nothing.

After returning to Rome following the completion of *Summa contra Gentiles,* Aquinas wrote several more commentaries on Aristotle. He then started what is regarded as his greatest work, the *Summa Theologiae* (Summary of Theology). In 1273, after suffering either a breakdown or some sort of life-changing spiritual experience, Aquinas refused to do any more writing. He would live for barely another year and died on March 7, 1274.

Pope John XXII began the process of canonization for Aquinas in 1316, and St. Thomas was sanctified in 1323.

ROGER BACON

Roger Bacon is an interesting figure in the history of science. More is claimed about him than historians know with certainty. Even his birth date is a matter of some speculation. Bacon was born in England between 1210 and 1219. He attended the University of Paris during the period 1237–1247. While there, Bacon likely associated with famed contemporaries such as the Dominican scholar Albertus Magnus (or Albert the Great). Bacon then returned to England and studied at Oxford until about 1257. He moved again to Paris around 1260 after joining the Franciscan order of friars, founded by St. Francis of Assisi. Franciscans embraced poverty and an itinerant life as preachers. Some members were highly learned, but most were ordinary people; their message was one of repentance. In 1257, the Franciscans forbade Bacon to teach, though the reason for this proscription is unknown. He did, however, continue to write, and, in 1266, Pope Clement IV requested that Bacon send him some writings. By 1277 or 1278, Bacon was condemned as a heretic because of his theological views (scholars believe that they were too liberal for the conservative Franciscans), and he spent the remainder of his life in prison until his death, in 1292.

Bacon began his career as a scholastic philosopher and produced an accomplished commentary on Aristotle; in addition, he was an early supporter of translations of the Greek philosopher. He lectured on Aristotelian philosophy at Paris, and some of these lectures survive today. However, Bacon is best known for his work on optics (the study of light and lenses). He produced a collection of contemporary writings on the subject that included works by both Christian and Arabic authors. Most scholars, including Albert the Great, supported Aristotle's understanding of vision, which was that sight resulted from alterations in an all-encompassing invisible medium caused by some object and the transmission of that alteration to the eye. The theories of the Muslim scholar Alhazen drew particular attention from Bacon, who endorsed Alhazen's view that an object gave off forms of itself in rays that interacted with the eye to produce vision. Studies of the rainbow, of refraction, and of the speed of light also occupied Bacon's intellectual energies.

Bacon was a consistent promoter of natural philosophy and a champion of educational reform that would emphasize the importance of that subject in the medieval university curriculum. In this way, the medieval Bacon was much like his seventeenth-century counterpart, Francis Bacon, whose calls for new methods in natural philosophy are often seen as the impetus for the founding of the Royal Society of London. In his writings, Roger Bacon emphasized experiment and mathematics when studying nature. His best-known work, *Opus Maius* (1266), challenged what Bacon viewed as the chief source of error in natural philosophy: blind acceptance of authority. *Opus Maius* (The Larger Work) outlines what Bacon viewed as impediments to proper wisdom and knowledge of nature. Bacon stated that there existed one true wisdom, which was to be found in the Bible; however, natural philosophy was also contained within this wisdom. God might be known through either scripture or His creation, the natural world. To arrive at an accurate understanding of God's creation and hence God Himself, Bacon argued that one needed the proper tools. Such tools included geometry, mathematics, and astronomy (French and Cunningham 1996, 238–243).

It was only through personal investigation that one came to real knowledge of nature, Bacon argued. To achieve this, Bacon offered two tools. First, one had to commence with dedicated study of languages so that any errors in translation, say from the Greek original of Aristotle to the Latin of the university, might be identified and corrected. Second, proficiency in mathematics was required because without it one could not perform proper astronomy. Bacon's insistence that mathematics be applied to the study of nature was unique in his day. Aristotle had explicitly denied that mathematics offered anything of use to natural philosophy, which for him consisted of contemplating the being and form of a thing, in addition to the various ways that the thing might change. Plato had depicted a universe that ran on geometrical principles, but, with the recovery of Aristotelian philosophy, around 1200, mathematics became separated from natural philosophy. For Bacon, mathematics was the easiest to learn of all the natural sciences; to master it was to prepare oneself for more detailed study of the world.

While Bacon is traditionally credited with the invention of a new kind of natural philosophy, called *scientia experimentalis* (experimental science), his was not a secular enterprise. All this was done so that natural philosophy might be put to the best use: the support of Christianity. As a further example, Elspeth Whitney notes that Bacon wished to make better lenses so that light might be focused into a weapon to destroy the enemies of God (Whitney 2004, 155).

SCHOLASTICISM

By the early thirteenth century, much of Aristotle's philosophical writings had been translated into Latin and was available to Christian scholars for the first time since the fall of the Roman Empire. These works included original works of Aristotle and Arabic commentaries on them that were included under the umbrella of the Aristotelian corpus. This rediscovery happened to coincide with the establishment of European universities. Typically, classes consisted of a master reading from the text of an accepted authority and the students dutifully copying what was said. By 1255, Aristotle was the agreed-upon authority, and his works formed the basis of medieval university education. Aspects of Aristotle discussed in the academic setting included philosophical and theological questions and those of a more cosmological nature.

Scholasticism describes both a literary form and a philosophical method. It originated in debates conducted within the university classroom and is rightly called the philosophy of schools: hence, scholasticism. While there were a variety of definitions, a "Scholastic" philosopher or "Schoolman" (the terms were often interchangeable) was a man who composed a commentary on Aristotle's work and who had studied at a university (usually Oxford or Paris), where he was likely to have held a teaching position and where he wrote his commentary.

There were, according to Edward Grant, a historian of medieval science, two approaches to teaching Aristotle's natural philosophy in a scholastic mode. First was the commentary. In this format the author would select some Aristotelian text, say *De Caelo* (On the Heavens), for example, and quote from selected passages before offering an interpretation of the quoted material. As a result, many scholastic works have a title that includes the phrase "Commentary on" a particular authority or text. Second was the question format. Here an author would pose inquiries and seek answers within Aristotelian philosophy or from authoritative commentators. Scholastic works began with grand questions or topic for discussion. The grand questions were divided into a series of articles, each of which focused on a particular issue within the question and took the form of a yes/no inquiry. Analysis began by expressing either agreement or disagreement with the position advanced in the article. Finally, the particular author of the scholastic work offered his answer to the question. Support for the views advanced came from the writings of Aristotle and from early Arabic commentators on Aristotle such as Avicenna and Averroes. Such questions included:

Whether the universe could have existed from eternity.
Whether the world will end at some time.
What was the light made by God on the first day of the world.
Why the production of stars was left to the fourth day.
Whether the world is perfect.
Whether Heaven is spherical in shape.
Whether the planets are spherical in shape. (Grant 1994)

Scholastic thinkers were not slavishly attached to the writings of acknowledged authorities. They did believe, as Sten Ebbesen reveals, that (1) authoritative authors had outlined correct principles for the disciplines; (2) authorities had divided logic and knowledge in a reasonable way and written works that addressed all the main topics; (3) the proper way to conduct philosophical inquiry was to follow the path set out by the authorities, careful never to contradict them but demonstrating what the authority meant to say in case of error; (4) Aristotle was the chief authority. The scholastic method may be traced to earlier scholarship done in areas where Greek was the language of scholarship from around 150–550 c.e. (Kretzmann, Kenny, and Pinborg 1982, 101).

Scholasticism and its Aristotelian support remained the dominant form of intellectual inquiry in medieval universities from around 1200 to the middle of the seventeenth century. During this 500-year period, the use of pure Aristotelean arguments gave way to an embrace of a modified Aristotelian approach by writers who still believed in an Earth-centered universe but who also incorporated the results of stellar observation into their scholastic works. Only with the mathematical demonstrations found in Isaac Newton's work on gravity and planetary motion was Aristotle's worldview completely overthrown and scholasticism abandoned.

CONDEMNATIONS OF 1277 AND AFTERMATH

By 1200, Aristotle was available in translation, and within 10 years the universities at Oxford and Paris were teaching his works. There seems to have been little worry about this in Oxford, but officials in Paris were concerned about the more overtly pagan (non-Christian) aspects of Aristotelian philosophy, such as the eternity of the world. In 1277, the Bishop of Paris, Etienne Tempier, condemned 219 articles of natural philosophy then being taught in universities throughout Europe. This event is known to historians as the Condemnation of 1277. While really applicable only to the University of Paris, the condemnations had a widespread effect. Some of the positions condemned included these: "That nothing is known better because of theology" and, which was sure to have irritated theologians, "That the only wise men of the world are philosophers."

Earlier attempts at banning part of Aristotle's works in and around the University of Paris had occurred between 1210 and 1230 but proved unsuccessful. In 1270, Etienne Tempier, already Bishop of Paris, condemned 13 positions in Aristotelian natural philosophy on the grounds that the positions were trespassing on the theological realm. Members of the arts faculty responded in 1272 by promising not to tackle questions of a theological bent in their lectures on natural philosophy, but they did not abandon the writings of Aristotle, who was known simply as "the philosopher."

The University of Paris had arguably both the best theological school and the best arts faculty in medieval Europe. Where the arts professors openly welcomed the Aristotelian corpus into their curriculum after the works had been recovered and translated in the thirteenth century, the doctors of divinity were suspicious of many natural philosophical positions advanced in the tracts. The condemnations reveal the deep divide in medieval universities between the faculties of arts and those of theology. The conflict was also about which method of knowing was best: reason or revelation. Disciplines in arts relied upon reason to make and judge philosophical arguments. Theologians deferred to revelation and faith when evaluating a particular piece of scripture. When sections of Aristotle's natural philosophy seemed to contradict what was known to be true through faith (for example, the eternal being of the world versus creation according to Genesis), conflict between the two faculties was certain. What is more, scholars in the faculty of theology saw themselves as the pinnacle of the university community and believed that the other faculties should be beneath them in terms of academic standing. Members of the arts faculty thought differently as they attempted to assert their independence within the university.

As historians have outlined it, the issue at stake was nothing short of God's power in the universe. The Bishop of Paris wished to emphasize God's absolute power (*potentia Dei absoluta*) with respect to the creation and with respect to the natural philosophical positions of Aristotle and Aristotle's commentators, which suggested that the workings of nature might restrain God's ability to act in the world. For example, Aristotle stated explicitly that a vacuum could never exist in the world and that other worlds like this one could not come into being. For theologians who praised the omnipotence of God, these rules could not stand. In their view, God could create any vacuum He wished. The fact that He had not yet done so did not mean that He lacked the power to do it.

For many philosophers writing after 1277, the solution was to include a proviso in their writings that said God had the ability to counteract the philosophical position they had just described. As an example, it became common to assert the absolute power of God, who was imagined to create all sorts of seemingly impossible spaces and vacuums. Another way to avoid problems was to say that the philosophical writing was hypothetical; to use the terminology of the time, the philosopher was said to be "speaking naturally" (Grant 1996). The effect of the Condemnation was not permanent. By 1325, some condemnations of articles had been revoked, and by 1341 lecturers at Paris were required to swear an oath to teach Aristotelian doctrine.

While this episode might be seen as the victory of religion over science, historians offer a more complicated account. Some argue that by being forced to embrace theoretical explanations that went beyond what Aristotle had written, medieval scholars found their minds opened to new possibilities. Indeed, Pierre Duhem, in the early twentieth century, stated that the origins of modern science should be dated to 1277, when philosophers were forced to find alternatives to Aristotle. Duhem's thesis found support in the work of the famed historian of science Lynn Thorndike, writing in the 1920 and 1930s, and, more recently, in studies by Edward Grant.

ASTROLOGY AND PORTENTS IN THE SKY

Astrology in medieval Europe attempted to explain how celestial bodies caused events and in some cases influenced people on Earth. Two points need to be made here. First, many respected medieval thinkers, and indeed the majority of everyday people, accepted the idea of stellar influence. Second, the study of this phenomenon was serious scholarship. There was not a uniform Catholic Church response to astrology. Some, including St. Augustine, rejected it, while the Dominican Albert the Great, among others, practiced some astrology as part of his natural philosophy. What is more, astrology was taught at Dominican schools around 1250. While there were certainly charlatans who were motivated only by money, others undertook a professional study of the sky to determine what the coming days and weeks held. This type of astrology, called judicial astrology, was somewhat controversial because it was thought that attributing human actions and choices to the stars and planets minimized the importance of both free will and God's power. This may have been St. Augustine's objection and the reason that during the early medieval period there was strong hostility toward astrology. Still, many royal courts employed astrologers to help make political decisions and to chart the course of state. Some doctors used astrology to determine the best time to administer cures. The more generally accepted astrology was concerned with accounting for the effects of the heavens, specifically the sun and the moon, on Earth itself in the form of storms, tides, and the like. In the 1130s, the translation of two ancient books into Latin sparked renewed interest in this kind of astrology. Ptolemy's *Tetrabiblos* and the Islamic writer Abu Ma'shar's *Greater Introduction to the Science of Astrology* provided astronomical data that supported astrological conclusions (Whitney 2004, 73).

This subject should not be dismissed out of hand. In modern times, we know that light and heat from the sun causes plants to grow and that the moon is connected to the tides; medieval natural philosophers understood this, too, but did not know exactly how. It was a very small leap to suggest that since the moon affected the seas, its influence might extend to other liquids, too. If the sun grew plants, other celestial bodies might affect the efficacy of medicine made from plants. For Albert the Great, light was the mechanism used by planets to exert influence. However, this was not inert light originating in lifeless planets; God Himself used the planets to propagate light as a providential tool. Thus, astrology was, for Albert, the study of God's action in the heavens, the means by which God chose to act in the universe.

For those who did not believe that astronomical bodies physically caused events, the appearance of specific objects in the sky was often seen as a sign that something extraordinary was about to happen. The medieval authority on astronomical events or portents (predictors or omens), the most influential of which was the appearance of comets, was the Greek philosopher Aristotle. In his *Meteorologica,* Aristotle claimed that comets were nothing more than appearances in the atmosphere caused by hot, dry air rising and interacting with the spheres of the heavens. While there were different types of comets, Aristotle remained convinced that they had no physical reality but were optical illusions only. Among those who concurred with Aristotle was Claudius Ptolemy, the Roman astronomer who provided medieval Europe with the mathematical tools needed to compute planetary orbits.

For Aristotle the appearance of a comet preceded more disturbing events such as strong winds, storms, earthquakes, and other natural forces. Anything that might be associated with an abundance of hot, dry air was often thought to follow a comet. The appearance of comets during important events secured their role as cosmological prognosticators of doom and fortune. For example, comets were seen at the time of the Achaen war, the assassination of Caesar, and the defeat of Anthony and Cleopatra's fleet.

During the medieval period, comets retained their role as heavenly signs but did so within a strongly providential Christian worldview. Church fathers who dismissed astrology nonetheless accepted the appearance of comets as a sure sign of some impending event, whether of divine or secular political occurrence. A comet was associated with predictions of defeat for King Harold at the hands of William the Conqueror in 1066. People in the countryside drew direct correlation between unusual events in the night sky and the potential for deaths in the village. A comet was also a warning from God, who dispatched them as a signal for sinners to repent before more serious divine powers were unleashed on unbelievers. Some medieval scholars argued that the most famous of all Christian astronomical signs—the star of Bethlehem—was a comet created by God to announce the birth of Christ. Comets, however, were not always good signs. St. Thomas Aquinas, citing the earlier St. Jerome, claimed that a comet would be one of the final indicators before the Apocalypse (Schechner 1997).

GLOSSARY

Ard: Early medieval plow that scratched the soil.

Arquebuses (Harquebusses): The first handgun; also a generic description for medieval firearms.

Ashlar Stone: Stones that are made smooth and square to be used in medieval building projects.

Astrolabe: Medieval astronomical tool used initially by Arabic scholars to calculate stellar positions.

Barrel Vault: Style of ceiling found in Romanesque churches; created by a sequence of rounded arches.

Bog Ore: Iron found in swamps and bogs that results from the oxidation of iron carried in the water.

Bombards: The largest early cannon.

Church Father: Early writer on Church doctrine and theology, usually from the second to the seventh century C.E.

Clinker Built: Method for building medieval ships that uses overlapping planks.

Codex: Earliest form of a book that would be recognized by modern eyes as a book.

Computus: The process of using mathematics and astronomy to determine a perpetual date for Easter. Also the written results of that process.

Deferent: The larger circle describing the overall orbit of a planet in Ptolemy's epicycle system.

Equant: The point about which a planet orbits, not necessarily the center.

Equinoxes: Days of equal hours of daylight and darkness.

Flying Buttress: External structural elements on Gothic cathedrals that allow the weight of the roof to be supported outside the cathedral walls.

Ecliptic: The hypothetical path traced by the sun in the heavens.

Epicycle: The smaller circle carrying a planet set upon a deferent in Ptolemy's epicycle system.

Gothic: A style of architecture famous for pointed arches and many stained-glass windows.

Humors: The four important fluids of the human body.

Natural Philosophy: The proper name for the study of nature in medieval Europe that emphasizes the world as God's creation.

Parchment: Medieval "paper" made from animal skin, most often from cows.

Pattern-Welding: A technique for producing swords that uses forged iron rods.

Ribbed Vault: Ceiling style, characteristic of Gothic design, that relies on intersecting supports.

Romanesque: An architectural design that is characterized by thick walls and rounded arches.

Scholasticism: Method of medieval intellectual inquiry that uses Aristotelian philosophy to answer philosophical questions.

Signatures: A group of pages that are sewn into a medieval codex.

Solstices: The shortest and longest days of the year.

Technological Determinism: The view that changes in technology are the driving force for change in society.

Trebuchet: A medieval catapult that achieves its power from the rotation of an off-center beam.

FURTHER READING

SOURCES CITED

Astill, Grenville, and Annie Grant, eds. 1988. *The Countryside of Medieval England.* Oxford: Basil Blackwell.

Astill, Grenville, and John Langdon, eds. 1997. *Medieval Farming and Technology: The Impact of Agricultural Change in Northwest Europe.* Leiden: Brill.

Avrin, Lelia. 1991. *Scribes, Script and Book: The Book Arts from Antiquity to the Renaissance.* Chicago: American Library Association.

Ayton, Andrew. 1994. *Knights and Warhorses: Military Service and the English Aristocracy under Edward III.* Woodbridge: The Boydell Press.

Bailey, Mark, ed. 2002. *The English Manor c. 1200–c. 1500: Selected Sources Translated and Annotated.* Manchester: Manchester University Press.

Balmer, R. T. 1978. "The Operation of Sand Clocks and Their Medieval Development." *Technology and Culture* 19: 615–32.

Battles, Matthew. 2003. *Library: An Unquiet History.* New York: W. W. Norton.

Beale, Philip. 1998. *A History of the Post in England from the Romans to the Stuarts.* Aldershot: Ashgate.

Beckingham, Charles F., and Bernard Hamilton, ed. 1996. *Prester John the Mongols and the Ten Lost Tribes.* Aldershot: Variorum.

Bede. 2004. *The Reckoning of Time,* trans. Faith Wallis. Rev. ed. Liverpool: Liverpool University Press.

Benedictow, Ole J. 2004. *The Black Death 1346–1353: The Complete History.* Woodbridge: The Boydell Press.

Berman, Constance Hoffman. 1986. *Medieval Agriculture, the Southern French Countryside, and the Early Cistercians: A Study of Forty-Three Monasteries.* Philadelphia: The American Philosophical Society.

Bischoff, Bernhard. 1994. *Manuscripts and Libraries in the Age of Charlemagne,* trans. Michael Gorman. Cambridge: Cambridge University Press.

Blackburn, Bonnie, and Leofranc Holford-Strevens. 1999. *The Oxford Companion to the Year.* Oxford: Oxford University Press.

Blake, John. 2004. *The Sea Chart: The Illustrated History of Nautical Maps and Navigational Charts.* London: Conway Maritime Press.

Boyer, Marjorie Nice. 1976. *Medieval French Bridges: A History.* Cambridge: The Medieval Academy of America.

Bradbury, Jim. 2004. *The Routledge Companion to Medieval Warfare.* New York: Routledge.

Brereton, J. M. 1976. *The Horse in War.* London: David and Charles.

Brisac, Catherine. 1986. *A Thousand Years of Stained Glass,* trans. Geoffrey Culverwell. Garden City: Doubleday.

Brown, George Hardin. 1987. *Bede the Venerable.* Boston: Twayne.

Brown, Sarah, and David O'Connor. 1991. *Medieval Craftsmen: Glass-Painters.* Toronto: University of Toronto Press.

Bruton, Eric. 1979. *The History of Clocks and Watches.* New York: Rizzoli International.

Bullough, Vern L. 2004. *Universities, Medicine and Science in the Medieval West.* Aldershot: Variorum.

Burke, James. 1978. *Connections.* Boston: Little, Brown.

Campbell, Mary B. 1988. *The Witness and the Other World: Exotic European Travel Writing, 400–1600.* Ithaca: Cornell University Press.

Coldstream, Nicola. 1991. *Medieval Craftsmen: Masons and Sculptors.* Toronto: University of Toronto Press.

Coldstream, Nicola. 2002. *Medieval Architecture, Oxford History of Art.* Oxford: Oxford University Press.

Cook, Olive. 1974. *The English Country House: An Art and a Way of Life.* London: Thames and Hudson.

Corson, Richard. 1967. *Fashion in Eyeglasses.* Chester Springs, PA: Dufour.

Courtenay, Lynn T. 1985. "Where Roof Meets Wall: Structural Innovations and Hammer-Beam Antecedents, 1150–1250." In *Science and Technology in Medieval Society,* ed. Pamela O. Long. New York: New York Academy of Sciences.

Crosby, Alfred W. 1997. *The Measure of Reality: Quantification and Western Society, 1250–1600.* Cambridge: Cambridge University Press.

Crosby, Alfred W. 2002. *Throwing Fire: Projectile Technology through History.* Cambridge: Cambridge University Press.

Crowfoot, Elisabeth, Frances Pritchard, and Kay Staniland. 2001. *Textiles and Clothing c. 1150–c. 1450.* Woodbridge: The Baydell Press.

Cunningham, Andrew. 1997. *The Anatomical Renaissance: The Resurrection of the Anatomical Projects of the Ancients.* Aldershot: Scolar Press.

De Hamel, Christopher. 1992. *Medieval Craftsmen: Scribes and Illuminators.* Toronto: University of Toronto Press.

DeVries, Kelly. 1990. "Military Surgical Practice and the Advent of Gunpowder Weaponry." *Canadian Bulletin of Medical History* 7: 131–146.

DeVries, Kelly. 1992. *Medieval Military Technology.* Peterborough: Broadview Press.

Dohrn-van Rossum, Gerhard. 1996. *History of the Hour: Clocks and Modern Temporal Orders,* trans. Thomas Dunlap. Chicago: University of Chicago Press.

Edge, David, and John Miles Paddock. 1988. *Arms and Armour of the Medieval Knight.* London: Bison Books.

Edson, Evelyn. 1999. *Mapping Time and Space: How Medieval Mapmakers Viewed Their World.* London: The British Library.

Eisenstein, Elizabeth L. 1983. *The Printing Revolution in Early Modern Europe.* Cambridge: Cambridge University Press.

Ffoulkes, Charles. [1912] 1988. *The Armourer and His Craft from the XIth to the XVIth Century.* Reprint, New York: Dover.

Flegg, Graham, ed. 1989. *Numbers through the Ages.* London: Macmillan.

French, Roger, and Andrew Cunningham. 1996. *Before Science: The Invention of the Friars' Natural Philosophy.* Aldershot: Ashgate.

Friedman, John Block. 1981. *The Monstrous Races in Medieval Thought and Art.* Cambridge, MA: Harvard University Press.

Garrison, Ervan. 1991. *A History of Engineering and Technology: Artful Methods.* Boston: CRC Press.

Geck, Elisabeth. 1968. *Johannes Gutenberg: From Lead Letter to the Computer.* Bad Godesberg: Inter Nationes.

Gimpel, Jean. 1976. *The Medieval Machine: The Industrial Revolution of the Middle Ages.* New York: Penguin Books.

Grant, Edward, ed. 1974. *A Source Book in Medieval Science.* Cambridge, MA: Harvard University Press.

Grant, Edward. 1994. *Planets, Stars, and Orbs: The Medieval Cosmos, 1200–1687.* Cambridge: Cambridge University Press.

Grant, Edward. 1996. *The Foundations of Modern Science in the Middle Ages: Their Religious, Institutional, and Intellectual Contexts.* Cambridge: Cambridge University Press.

Graves, Rolande. 2001. *Born to Procreate: Women and Childbirth in France from the Middle Ages to the Eighteenth Century.* New York: Peter Lang.

Green, Monica H. 1989. "Women's Medical Practice and Health Care in Medieval Europe." In *Sisters and Workers in the Middle Ages,* ed. J. Bennett et al. Chicago: University of Chicago Press.

Gregory, Cedric E. 1980. *A Concise History of Mining.* New York: Pergamon Press.

Griffiths, Jeremy, and Derek Pearsall, eds. 1989. *Book Production and Publishing in Britain 1375–1475.* Cambridge: Cambridge University Press.

Hall, Bert S. 1997. *Weapons and Warfare in Renaissance Europe: Gunpowder, Technology, and Tactics.* Baltimore: The Johns Hopkins University Press.

Harris, Michael H. 1995. *History of Libraries in the Western World.* 4th ed. London: The Scarecrow Press.

Harrison, David. 2004. *The Bridges of Medieval England: Transport and Society 400– 1800.* Oxford: The Clarendon Press.

Hill, Donald. 1984. *A History of Engineering in Classical and Medieval Times.* La Salle: Open Court.

Holt, Richard. 1988. *The Mills of Medieval England.* Oxford: Basil Blackwell.

Horrox, Rosemary, ed. 1994. *The Black Death.* Manchester: Manchester University Press.

Hughes, Quentin. 1974. *Military Architecture.* London: Hugh Evelyn.

Hunt, Edwin S., and James M. Murray. 1999. *A History of Business in Medieval Europe, 1200–1550.* Cambridge: Cambridge University Press.

Hutchinson, Gillian. 1994. *Medieval Ships and Shipping.* Rutherford: Fairleigh Dickinson University Press.

Johns, Adrian. 1998. *The Nature of the Book: Print and Knowledge in the Making.* Chicago: University of Chicago Press.

Jonkers, A.R.T. 2003. *Earth's Magnetism in the Age of Sail.* Baltimore: The Johns Hopkins University Press.

Kemp, Wolfgang. 1997. *The Narratives of Gothic Stained Glass,* trans. Caroline Dobson Saltzwedel. Cambridge: Cambridge University Press.

Kenny, Anthony. 2005. *A New History of Western Philosophy.* Vol. 2, *Medieval Philosophy.* Oxford: Clarendon Press.

Kenyon, John R. 1990. *Medieval Fortifications.* New York: St. Martin's Press.

Kilgour, Frederick G. 1998. *The Evolution of the Book.* Oxford: Oxford University Press.

Kretzmann, Norman, Anthony Kenny, and Jan Pinborg, eds. 1982. *The Cambridge History of Later Medieval Philosophy.* Cambridge: Cambridge University Press.

Krirshner, Julius, ed. 1974. *Business, Banking, and Economic Thought in Late Medieval and Early Modern Europe: Selected Studies of Raymond De Roover.* Chicago: University of Chicago Press.

Kuhn, Thomas. 1957. *The Copernican Revolution: Planetary Astronomy in the Development of Western Thought.* Cambridge, MA: Harvard University Press.

Landels, John G. 1979. "Water-Clocks and Time Measurement in Classical Antiquity." *Endeavour* 3 (1): 32–37.

Landes, David S. 1983. *Revolution in Time: Clocks and the Making of the Modern World.* Cambridge, MA: The Belknap Press of the Harvard University Press.

Langdon, John. 1986. *Horses, Oxen and Technological Innovation: The Use of Draught Animals in English Farming from 1066–1500.* Cambridge: Cambridge University Press.

Langdon, John. 2004. *Mills in the Medieval Economy: England 1300–1540.* Oxford: Oxford University Press.

Lay, M. G. 1992. *Ways of the World: A History of the World's Roads and of the Vehicles That Used Them.* New Brunswick: Rutgers University Press.

Le Goff, Jacques. 1988. *Medieval Civilization 400–1500,* trans. Julia Barrow. Oxford: Basil Blackwell.

Leighton, Albert C. 1972. *Transportation and Communication in Early Medieval Europe AD 500–1100.* Devon: David and Charles.

Lepage, Jean-Denis G. G. 2002. *Castles and Fortified Cities of Medieval Europe: An Illustrated History.* Jefferson, NC: McFarland.

Lindberg, David C. 1992. *The Beginnings of Western Science: The European Scientific Tradition in Philosophical, Religious, and Institutional Context, 600 B.C. to A.D. 1450.* Chicago: University of Chicago Press.

Lindberg, David C. 1995. "Medieval Science and Its Religious Context." *Osiris* 10: 61–79.

Lozovsky, Natalia. 2000. *'The Earth Is Our Book': Geographical Knowledge in the Latin West ca. 400–1000.* Ann Arbor: The University of Michigan Press.

Mark, Robert, and Huang Yun-Sheng. 1985. "High Gothic Structural Development: The Pinnacles of Reims Cathedral." In *Science and Technology in Medieval Society,* ed. Pamela O. Long. New York: New York Academy of Sciences.

Marks, P.J.M. 1998. *The British Library Guide to Bookbinding: History and Techniques.* Toronto: University of Toronto Press.

Martin, Geoffrey J. 2005. *All Possible Worlds: A History of Geographical Ideas.* 4th ed. Oxford: Oxford University Press.

McClellan, James E. III and Harold Dorn. 1999. *Science and Technology in World History: An Introduction.* Baltimore: The Johns Hopkins University Press.

McCluskey, Stephen C. 1998. *Astronomies and Cultures in Early Medieval Europe.* Cambridge: Cambridge University Press.

McGowan, Alan. 1981. *The Ship: Tiller and Whipstaff: The Development of the Sailing Ship 1400–1700.* London: Her Majesty's Stationery Office.

McKitterick, Rosamond. 2001. *The Early Middle Ages.* Oxford: Oxford University Press.

Menache, Sophia. 1990. *The Vox Dei: Communication in the Middle Ages.* Oxford: Oxford University Press.

Mueller, Reinhold C. 1997. *The Venetian Money Market: Banks, Panics, and the Public Debt, 1200–1500.* Baltimore: The Johns Hopkins University Press.

Musacchio, Jacqueline Marie. 1999. *The Art and Ritual of Childbirth in Renaissance Italy.* New Haven: Yale University Press.

Nauert, Charles G. Jr. 1995. *Humanism and the Culture of Renaissance Europe.* Cambridge: Cambridge University Press.

Nicholson, Helen. 2004. *Medieval Warfare: Theory and Practice of War in Europe 300–1500.* New York: Palgrave Macmillan.

Osler, Margaret J. 1997. "Mixing Metaphors: Science and Religion or Natural Philosophy and Theology in Early Modern Europe." *History of Science* 35: 91–113.

Ovitt, George Jr. 1987. *The Restoration of Perfection: Labor and Technology in Medieval Culture.* New Brunswick: Rutgers University Press.

Pfaffenbichler, Matthias. 1992. *Medieval Craftsmen: Armourers.* Toronto: University of Toronto Press.

Phillips, W. D., and C. R. Phillips. 1992. *The Worlds of Christopher Columbus.* Cambridge: Cambridge University Press.

Piggott, Stuart. 1992. *Wagons, Chariot and Carriage: Symbols and Status in the History of Transport.* London: Thames and Hudson.

Porter, Roy. 1997. *The Greatest Benefit to Mankind: A Medical History of Humanity.* New York: W. W. Norton.

Radding, Charles M., and William W. Clark. 1992. *Medieval Architecture Medieval Learning: Builders and Masters in the Age of Romanesque and Gothic.* New Haven: Yale University Press.

Rawcliffe, Carole. 1997. *Medicine and Society in Later Medieval England.* Stroud: Sutton.

Reynolds, Susan. 1994. *Fiefs and Vassals: The Medieval Evidence Reinterpreted.* Oxford: Oxford University Press.

Rist, John M. 1994. *Augustine: Ancient Thought Baptized.* Cambridge: Cambridge University Press.

Rösener, Werner. 1992. *Peasants in the Middle Ages,* trans. Alexander Stützer. Cambridge: Polity Press.

Santosuosso, Antonio. 2004. *Barbarians, Marauders, and Infidels: The Ways of Medieval Warfare.* Boulder: Westview Press.

Schechner, Sara J. 1997. *Comets, Popular Culture, and the Birth of Modern Cosmology.* Princeton: Princeton University Press.

Schofield, Phillipp R. 2003. *Peasant and Community in Medieval England 1200–1500.* New York: Palgrave.

Short, John Rennie. 2003. *The World through Maps: A History of Cartography.* Toronto: Firefly Books.

Singman, Jeffrey L. 1999. *Daily Life in Medieval Europe.* Westport, CT: Greenwood Press.

Siraisi, Nancy G. 1990. *Medieval and Early Renaissance Medicine: An Introduction to Knowledge and Practice.* Chicago: University of Chicago Press.

Szirmai, J. A. 1999. *The Archaeology of Medieval Bookbinding.* Aldershot: Ashgate.

Tait, Hugh, ed. 2004. *Five Thousand Years of Glass.* Philadelphia: University of Pennsylvania Press.

Unger, Richard W. 1980. *The Ship in the Medieval Economy 600–1600.* Montreal: McGill-Queen's University Press.

Unger, Richard W. 2004. *Beer in the Middle Ages and the Renaissance.* Philadelphia: University of Pennsylvania Press.

Unwin, Tim. 1991. *Wine and the Vine: An Historical Geography of Viticulture and the Wine Trade.* New York: Routledge.

Verdon, Jean. 2003. *Travel in the Middle Ages,* trans. George Holoch. Notre Dame: University of Notre Dame Press.

von Simson, Otto. 1988. *The Gothic Cathedral: Origins of Gothic Architecture and the Medieval Concept of Order.* 3rd ed. Princeton: Princeton University Press.

Wade, Nicholas J. 1998. *A Natural History of Vision.* Cambridge: The MIT Press.

Wear, Andrew. 1995. "Medicine in Early Modern Europe, 1500–1700." In *The Western Medical Tradition 800 BC to AD 1800,* ed. Lawrence I. Conrad et al. Cambridge: Cambridge University Press.

White, Lynn Jr. 1962. *Medieval Technology and Social Change.* Oxford: Oxford University Press.

Whitney, Elspeth. 2004. *Medieval Science and Technology.* Westport, CT: Greenwood Press.

Williams, David. 1997. *Late Saxon Stirrup-Strap Mounts.* York: Council for British Archaeology.

Woolgar, C. M. 1999. *The Great Household in Late Medieval England.* New Haven: Yale University Press.

OTHER USEFUL SOURCES

Books

Day, Lance. 1998. *Biographical Dictionary of the History of Technology.* New York: Routledge.

Ede, Andrew, and Lesley B. Cormack. 2004. *A History of Science in Society: From Philosophy to Utility.* Peterborough: Broadview Press.

Fourace, Paul, et al., eds. 2005. *The New Cambridge Medieval History.* 7 vols. Cambridge: Cambridge University Press.

Glick, Thomas, et al., eds. 2005. *Medieval Science, Technology, and Medicine: An Encyclopedia.* New York: Routledge.

Grant, Edward. 2004. *Science and Religion, 400B.C.–A.D. 1550: From Aristotle to Copernicus.* Westport, CT: Greenwood Press.

Greenwood Guides to Historic Events of the Medieval World. (Twelve volumes thus far in this valuable series of reference books.)

McNeil, Ian. 1990. *An Encyclopaedia of the History of Technology.* New York: Routledge.

Singer, Charles Joseph. 1954. *A History of Technology.* 8 vols. Oxford: The Clarendon Press. Volumes 2 and 3 deal with medieval topics.

Smith, Elizabeth Bradford, ed. 1997. *Technology and Resource Use in Medieval Europe: Cathedrals, Mills, and Mines.* Aldershot: Ashgate.

Scholarly Journals

British Journal for the History of Science
Early Medieval Europe
Early Science and Medicine
History of Science
History of Technology
Isis
Journal of Medieval History
The Medieval History Journal
Technology and Culture
Traditio

Videos and Web Sites

"Castles and Dungeons." The History Channel, A&E Television Networks. New York. Available on DVD.

"Connections." James Burke narrates this entertaining 10-part television series that details the history of technology and its societal implications. It was originally broadcast on BBC in England and on PBS in America between 1976 and 1978. "Connections II" and "Connections III" aired on the Discovery Channel between 1995 and 1998. Available on VHS and DVD.

"The Medieval Soldier." A&E Television, A&E Television Networks. New York. Available on DVD.

"The Plague." The History Channel, A&E Television Networks. New York. Available on DVD.

"Worst Jobs in History." Six-part British television series, produced by Channel 4 Network, that re-creates some of the most unpleasant ways to earn a living throughout English history. It is not yet available (as of 2006) on DVD but airs regularly on the Discovery Channel in North America.

AVISTA, an organization that emphasizes study of art, science, and technology of the medieval era, has a Web site, available at http://www.avista.org.

The British Library has an extensive on-line image catalogue with many medieval pictures; available at http://www.bl.uk.

The History of Science Society Web site is available at http://www.hssonline.org.

James Buke's Knowledge Web is available at http://www.k-web.org.

"The Medieval Science Page" is a portal to a variety of Web pages dealing with medieval topics in science and technology; available at http://members.aol.com/mcnelis/medsci_index.html.

The Morgan Library Web site has many illustrations from medieval manuscripts; available at http://www.morganlibrary.org.

The Millersville University Web site deals with early exploration; it is available at http://muweb.millersville.edu/~columbus/.

The Society for the History of Technology Web site is available at http://shot.jhu.edu.

INDEX

About the Author

JEFFREY R. WIGELSWORTH is Scholar in Residence at the Calgary Institute for the Humanities. Dr. Wigelsworth has taught European history and history of science at the University of Saskatchewan, University of Calgary, and Mount Royal College, Calgary. His articles have appeared in *Isis, Canadian Journal of History, Journal of the Printing Historical Society,* and other venues.